Industrial Biotechnology Commercialization Handbook

*How to make proteins without animals and fuels
or chemicals without crude oil*

Mark Warner, PE

ISBN # 9781079229950

Cover Design by Isabel Paner

To Anne, for everything, including her unwavering support to make the world a better place for the next generation.

Table of Contents

Forward

There have been many great publications on the call to arms to bring novel proteins, alternative fuels and renewable chemicals to market. These are based a mix of animal rights and climate change concerns, to make the world a better and more sustainable place. This effort has rallied significant support and seen the rise of companies who have experienced a dramatic increase in valuation and investor interest. This begs the question, now what? There is significant interest, but what is the tactical plan to bring this vision to reality?

This handbook is intended to serve as a roadmap to successfully commercializing these technologies, based on lessons of the past, mixed with significant recent experience with many of the industrial biotechnology industry leaders and will serve as a useful resource guide for investors, board members, executives and employees all striving to make their novel biotechnology a commercial success.

Acknowledgments

This handbook is a compilation of many years of white papers and presentations on a range of topics in advanced biotechnology commercialization. While I authored all these documents, the work products all benefitted significantly from my experience at leading biotechnology and engineering companies and learning from the most talented people in the industry. It was this time and the things I learned working side-by-side with a broad range of experiences that I hope are reflected in the handbook. I believe I am better for those experiences and hope it reflects in the work product.

I wanted to make a special acknowledgment to Jim Lane and Biofuels Digest for being my primary venue to publish many of the documents compiled here. The broad and open access that Biofuels Digest provides to people commercializing advanced biotechnology is unique and greatly appreciated.

TOP TEN LESSONS LEARNED COMMERCIALIZING ADVANCED BIOTECHNOLOGIES

Spending half a billion dollars placing steel in the ground to commercialize advanced biotechnologies for fuels, chemicals and food applications, is the easy part. Doing it successfully is the hard part – but I have been fortunate to lead many successful projects over the last ten years, and that is what has prompted me to share my personal experience. I have done my best to include real world examples and attributes that make ventures succeed, as well as common pitfalls. My focus is on the technical perspectives of technology deployment, from early stage process development, engineering, and construction to a fully operational commercial facility. This handbook is for anyone who is passionate about scaling advanced biotechnology dreams.

1. Building the Team – *Capability, agility, and diversity are critical*

The main lesson I learned, the hard way, is there is not a right or wrong type of experience, it comes down to the capability of the team to transform as broader challenges emerge. The toughest challenges are rarely in proving the technology – it's in scaling and commercializing it. This is a rapid-deployment game, won only by the team whose members have the capacity to listen and learn, the agility to quickly transition, and the desire to embrace diverse backgrounds who will take informed risks to succeed.

As a startup grows and matures, it is inevitable that broader challenges will emerge, many that are not related to the core technology. This often happens when the initial focus of the company is scientific discovery and it transitions to commercial deployment. While a complex scientific problem may be of little to no concern for a senior scientist, determining how that problem can be solved economically at commercial scale, including impacts that run the gamut from site selection to marketing, is problematic if the team is not fortified by those who have the skill and will to successfully transform to commercial stages. The scientist cannot be successful without partnering with industrial experience, and vice versa – it is the skill of the team that trumps all. The key is to identify the experience types that cover the range of challenges faced during the lifecycle of commercialization, and to hire individuals who respect that handoffs to others is never in doubt.

The ability to identify when an employee is not a fit for advanced deployment or worse, destructive to the culture, is key. Often, there is a feeling that the specific technical experience the employee brings is critical, but the real test is a leader's capacity to transform the team and the employee's ability to understand the larger purpose which will evolve from its original concept in the lab.

Harmonizing the need for exceptional technical staff who can evolve through rapid growth stages of a startup venture, and transform to long-term key contributors during commercialization, is the key to ultimate success.

2. Process Development – *It's not all about you*

When looking at the long-term challenges that manifest during developing and deploying innovative biotechnologies, it is important to first understand how most biotechnologies are discovered and mature as the venture grows. Most new technologies emerge as a new pathway that is "light-years" ahead of the way things are currently done. The traditional model in the advanced bioeconomy is an organism that can produce a desired compound more efficiently, sustainably, and (hopefully) cost effectively than what is in commerce today.

For our example, let's assume we have discovered a wonder bug that is able to produce our target compound, di-methyl glop (DMG), very efficiently via fermentation. DMG is used in many specialty chemical and commodity applications, and we are confident that if we hit the price projections, we will have an industry-changing venture. We use the positive discovery to form a startup company, get VC funding from a handful of name venture funds and build a world-class team of scientists, with a robust toolkit of genetics and process technology to improve our yield and productivity to the point our fermentation process can be commercially viable. Sounds great, right? It is, but unfortunately, it's only the first leg on the long journey to commercialization. What has been made is a wonderful fermentation broth rich in very valuable DMG, but this now has to be separated and refined to a point it meets industry specifications. It is that second portion of the process that is often the biggest challenge, not because it's necessarily harder, but because it is such a departure from the initial mission of the startup and the organization can struggle to make the shift.

The source of comfort with focusing on the core technology comes during the early development days at a startup, when yields and titers are tracked daily, new versions of the organism are cheered, and fermentation results are dutifully reported at every board meeting. The justified pride is such a motivation, the board presentation often includes analytical results only hours old. So, what is the problem? As wonderful as the technology is, it will only end up being a single cog in the overall machine that is a fully integrated production process. It is like being a proud parent of a star athlete, you know that your child is the best player on the team, but for the team to win, it may be more important that your child play his or her role as a key member of the integrated team, than be the individual MVP. The same is true for technology. If your ultimate goal is the best overall process for cost and quality of production, it may be more important that your core technology support the overall process, not be the individual star.

3. Scale-up - *There is no substitute for a fully integrated pilot process*

How a process is scaled-up and what constitutes an acceptable risk, to ensure your larger-scale facility operates as predicted, are some of the least appreciated concepts within the advanced bioeconomy. That said, both receive significant focus as companies move into larger-scale production and more importantly, as investors seek more certainty on the probability of a return on their investment.

First, let's start with the basics. It is important to realize the difference between scaling a unit operation and scaling an integrated process. The unit operations are components of the overall process and scaling them individually will seldom result in a properly scaled integrated process. Let's go back to our example of DMG. Our example process consists of four steps:

1. Aerobic fermentation to make DMG
2. Breaking the cells to release DMG into the broth and removing the spent cells
3. Concentration of DMG by multi-stage evaporation
4. Purification of DMG by crystallization

In this process, it is likely that steps #1 and #2 would be done internally at the company pilot plant with fermentation, homogenization and centrifuge equipment. The initial evaporation and crystallization efforts to prove the concept would likely be done at third party vendors. Shipping the purified broth to the evaporation vendor's pilot plant, then shipping that the concentrate to the crystallization vendor. This results in three different portions of the process development. Assuming all three of these individual pilot tests go as planned, generate engineering data (flows, temperatures, concentrations, etc.) and produce in-specification DMG, does this provide "proof of concept"? Absolutely. Is it adequate to scale to a larger facility? No, and here is why.

Proof of concept is very important in the development process, but it only shows that the process can work. It seldom provides enough detail to predict how it will work as an integrated process, taking into consideration items like recycle and byproduct streams and contaminants that may build up over time when recycled. This is why running an integrated process is an indispensable stage in process scale-up. It is when each unit operation is run in series, being interdependent upon each other, that scale-up challenges manifest themselves. Once proof of concept exists, the next critical step is to pilot-scale equipment to bring the concentration and purification steps in house. A general rule of thumb is 500 – 1,000 hours of integrated pilot plant data is a key indicator of readiness for the next level, with a reasonable expectation of success. The jump in scale should not be greater than 10 times the capacity of the unit operations tested in the pilot.

Now you are thinking to yourself, 500 - 1,000 hours of operation and I can only go up 10 times in scale, do I always need to do that? The answer is usually yes, but not always. If you are handling complex mixture with solids and liquid materials (like the DMG broth), it would be high risk to go outside of these guidelines. The case where being more aggressive in hours of operation and scale-up factor is when you are dealing primarily with fairly pure liquids and gasses. Much of the scale-up risk on processes comes around material handling and obtaining representative data from samples. In the case of liquids and gasses with a limited number of components, it's much easier. Regardless, there is no substitute for a fully integrated pilot process.

4. Facility Design – *Understanding fit for purpose*

Without a doubt, the largest source of technical difference of opinion, in the commercialization of a new bioprocess, is the design standard for building a larger-scale facility. The magnitude of the difference of opinion is directly correlated to whether the team members have the capacity to transform and hail from diverse backgrounds (as noted in Lesson #1). Everyone comes with experience and perspective, as well as biases for equipment and vendors, which may or may not be right for the process under development. To understand the source of the conflict, we need to look at what are the most common design standards and how we determine which one is fit for our purpose.

The birth of modern biotechnology began predominantly in the pharmaceutical industry, where the products being made have a very high value and the purity requirements are very strict. Many of these same technologies have been used in the emerging industrial biotechnology in-dustry (biofuels and biochemical), with more recent application to food products. As we look to build large-scale facilities, we will find that the standards (and resulting capital costs) can vary dramatically between each of these platforms. Let's take a look at the requirements for each and what standards would be used to design a fermentation and downstream processing system.

Pharmaceutical Standards – Fermentation used to support pharmaceutical production is usually referred to as sterile or aseptic, meaning that engineering controls are in place to ensure that the process is made completely sterile before fermentation and remains that way throughout. This means ending a fermentation with only your target organism and no other organisms (aka "a clean batch") is the expectation and typically the standard for most pharmaceutical applications.

The two main techniques that are used to ensure sterility are steam sterilization of the sys-tems prior to use and robust engineering techniques to ensure no viable foreign organisms can enter the system during operation or build up over time. There is a very specific ASTM standard that outlines these requirements. Additionally, there are strict rules on equipment validation that involve a bureaucratic amount of documentation, certification, and testing.

One easily identifiable attribute of this standard is all process surfaces end up having a near mirror finish. If you look inside a fermenter for a pharmaceutical application, you can usually see your reflection. This is done to ensure small organisms cannot find a surface to adhere and avoid sterilization. Downstream processing systems rely on similar technologies, including disposable systems. There is a trend toward using disposables from fermentation, through downstream processing, to avoid the litany of documentation and testing required to validate equipment for use. The cost to sterilize many types of equipment has made it more cost effective for simply replace versus sterilize.

Industrial Biotechnology Standards – As scale grows in size and output of production, it is not cost effective to build processes to pharmaceutical standards. It is also not usually required, as it is more about controlling the levels and types of other organisms, not their existence. The Pharmaceutical ASTM standard is often used as a guide, but it is adjusted to fit the proposed process. What is standard for a 250-liter packaged aseptic fermenter is not practical or cost effective for a 500,000-liter industrial fermenter. While there is always a desire for a completely clean batch (no competing organisms), it is not usually a requirement in industrial processes. This has generated an industrial biotechnology design standard that takes the basic concepts of high sterility fermentation, but based on risk factors, applies only the portions that are required. In the case of fermenter finish, these seldom are high-polish in large-scale applications, usually a standard milled stainless steel. Large-scale fabricators are often not set up to be able to polish on a large-scale and it can add very significant capital costs.

Food production standards – While the progression from pharmaceutical to industrial biotech was basically a "tweak" or re-application of similar standards, the recent move within the advanced bioeconomy to food applications has brought forward significant design challenges. The basic design concept within food is to ensure there are no pathogens (bad bugs), then limit and control the overall organism count. Food is not produced to be free of organism, just free of certain bad organisms and below acceptable levels of others.

While the basic concept of the first two standards was sterilization (ability to kill all organisms), food processing is based on the concept of cleanable. Many of the types of equipment used in food processing cannot be steam sterilized, so the applicable standards require it to be completely disassembled and cleaned on a regular basis. This is often quite surprising to technical staff who have not spent time in the food industry.

Comparing these three standards and having experience building and operating facilities under all 3 standards, the most important concept I have learned is determining what is fit for purpose. Just because a piece of equipment is designed to a high standard and is more expensive, does not mean is the right equipment for your application. I have seen many cases where very expensive pharmaceutical grade equipment is purchased for a food application and not only was orders of magnitude more expensive than the proper food equipment, it actually did not work as well. The challenge is for staff to rely less on their experience from other industries and be open to learning the needs and applying the appropriate standards of the target industry.

5. Integrating Standard Unit Operations - *Standard industrial processes are never standard*

This lesson has a special place in my heart, as it is a lesson I have learned the hard way - twice. You generate the core process of your technology, in the case of DMG, fermentation and cell separation, and you look at something like multi-stage evaporation and say to yourself:

- It is done all the time, there is nothing new here
- Our stream is very similar to other streams that use this all the time
- The process is low risk, so we do not need to assign many resources to it

If you have similar thoughts, please don't and here is why. While the technology may be proven and have a long history, it is for a certain type of material. Your feedstock may look similar to you, but when you get into the details, you will find it often is not. For multi-effect evaporation, there could easily be fouling or corrosion issues that can only be identified through fully integrated pilot testing.

The most common example of overconfidence on application of a standard industry technology I have seen is purification systems for oils. This technology is commonly referred to as refining, bleaching and deodorizing (RBD) and used interchangeably in the vegetable oil industry with great success. This relies on the fact that one type of natural oil (canola, soy, etc.) is similar to another and there is enough history on all of them for vendors to make the minor modifications needed. Given the long history and experience on vegetable oils, this has been very successful. The problem comes when applying these technologies to new streams that appear the same, but really are not.

Oils generated from both phototrophic and heterotrophic algae look similar to standard vegetable oils, except when you get to trace contaminants. In the case of phototropic algae there are often various salts and algae made by fermentation have residual sugars and fermentation media. These can be very different than the standard contaminants RBD is set up to deal with.   This is one more example of why there is no substitute for fully integrated pilot operations.

6. Site Selection – *Understand the implications of where you decide to build*

Deciding on where to build the plant is driven by commercial availability and economic considerations, such a simple statement – yet mired in failure points that require mitigation plans that only a Type-A personality can appreciate. Focus here is on ensuring all understand the risks and challenges up front and are capable of quickly enabling mitigation plans when the inevitable occurs.

Deploying a new technology in the United States is almost always preferable from a project delivery perspective. The ability to source equipment, staff with qualified employees and secure a quality construction contractor are orders of magnitude higher in the US than many parts of the world. As for the specific location your venture selects, that will depend on a handful of factors. Feedstock availability, utilities and proximity to existing operations are typically some of the driving factors. It is important to keep in mind the ability to grow and attract quality technical staff. In many cases, a small idled site in the rural Midwest may make a great demonstration plant, but if the intention is to expand into an eventual commercial operation, availability of utilities and difficulties convincing key technical staff to relocate may inhibit the expansion process. These are examples of risks to consider up front.

There has been a considerable push in recent years to build demonstration and commercial-scale biotechnologies outside of the US, with a primary focus on South America and Asia. These are prime examples of where feedstock supply and funding sources make deployment to these areas very attractive. In those cases, it is then important to understand what the implications are. Beyond the obvious concerns of culture and language, my main lesson learned is that it is much harder than you ever dreamed, for reasons never conceived. It doesn't mean it can't be done, you just need to go in prepared for the level of complexity it will bring. My main takeaways have been:

- It is not just a matter of being in a foreign country, but how rural the site location is. Building a major factory in a rural part of South America or Asia is more difficult to get vendors, equipment and key staff; even housing is a challenge. Be prepared for this as part of the commercialization process.
- Major projects cannot be managed remotely, you need boots on the ground throughout the project. This comes in two forms. You will always need senior in-country staff who are native to the country with a network. Do this up front. The second type of boots on the ground is key technology and engineering staff from your company. There will be hundreds of decisions to be made on a daily basis that cannot be done by phone or skype (if those even happen to be working that day from your remote location in a foreign country).
- Predict the challenges and plan for them. The key to successfully constructing and operating a major facility, especially in a foreign country, is to have a fully committed

technical team that understands the challenges and execution plan up front. Finding out late in the process you may need to spend months during startup out of the country does not usually end well, these expectations need to be clearly articulated up front when hiring and staffing the project.

- Hire early and often. It is inevitable you will get attrition from staff that can no longer spend extended time out of the country, and you need to have backups ready. This is one of those hidden costs of commercializing overseas. It is critical that there is a fully capable local staff that can handle plant operations and technology transfer to allow your startup team to return home.

I do not mean to infer that projects cannot and should not be done in a foreign country, there are many good commercial reasons. First-hand experience does oblige me to point out there needs to be very compelling reasons though. When in doubt, target the US for all initial operations. If there is a deal you cannot refuse in a foreign country, that's great, but go in with both eyes open.

7. Regulatory – *Know what you need to do from day one*

Nobody likes surprises, least of which are regulatory surprises that come after you have started commercializing your technology. Getting a good regulatory understanding in the early stages of development is critical. This is driven by the fact that most of the U.S. product regulations were developed over 30 years ago and did not envision the vast expanse of bio-based products. Just because your product is chemically identical to something on the market, does not mean it is regulated the same as the other product. It is unfair and often unreasonable, but unfortunately is a reality that needs to be dealt with.

Given this reality, get advice from someone experienced in the process, sometimes it may be more than one person. The larger regulatory concern in the advanced bioeconomy is whether the organism being used is regulated. If it is genetically modified, the answer is yes. The harder question to answer is what does that mean to your process? This is usually a complex determination, based significantly on how "new" the organism is and whether the organism is a tool to make a product or if it is the final product for a food application. If it is an organism being used to make products in commerce today, the path will be fairly clear. If it is not, then the process required and time needed may not be clearly defined. The key determinations that will come from this process are whether the products can be sold into commerce and what engineering controls are needed to contain the organism. These engineering controls can vary dramatically based on the risk that is determined to be posed by the organism and it is critical this is a design consideration that is known up front. Making changes late in the project can have a significant impact on schedule and cost.

This is a situation where it is best to seek a battle tested veteran who has been through the process before. Look for an independent consultant or often a retired executive who has been through the entire process. The reason for this is there is usually not a clear yes or no answer, and your company will need to make risk-based decisions. First-hand experience is very valuable in this situation. The best source of this is someone who has been through it many times. In the overall scope of the project, it is not that much money, and is penny wise.

8. Financing - *The tail does wag the dog*

The manner in which a project is financed has a dramatic impact on how the project can and needs to be executed. In the case of equity funding, the key decisions are with the company and projects can move very rapidly. The proof of concept of the technology and responsibility for key decisions, like purchasing major equipment, does not require outside approval. This is generally the gold standard of project funding from the engineering perspective and allows for the fastest project delivery. Unfortunately, it is not that common, especially for large expensive facilities, as it creates dilution for existing shareholders if significant capital is raised as equity.

The other end of the spectrum is traditional project finance where a bank loans a large portion of the money, but has significant restrictions and approvals. They will likely require a third-party assessment of the technology, review of all contracts for feedstock supply and product offtake, and require sign off on all major equipment purchases. As most new technologies do not have all these details locked down, traditional project finance is not common in new technologies. If used, the project timeline will need to reflect all these approvals and will need to be extended.

The most common form of financing for new technologies is a hybrid of equity, teamed with either a federal grant or a federally backed loan guarantees. In the case of grant, it is a source of funding that does not need to be paid back, but it is subject to meeting a series of technical hurdles. Loan guarantees look a lot like a traditional project finance, but the government accepts the technology risk and backs the loan. This streamlines the approval steps and control some, but does not remove it.

The decision on type of funding is usually outside of the technical team, but the important part is that the implications are understood. Time and effort can be wasted if the project delivery approach is not consistent with the requirements of the funding source.

9. Facility Start-up – *New facilities do not turn on like a light bulb*

Nothing sends chills down my spine like seeing the "step function" production forecasts for new technologies. A date is picked for startup of the new facility and the financial modeling shows it going from no production to near capacity almost immediately. This is unfortunately more common than most would believe and is often based on a limited understanding of what is involved in bringing a new facility on line, especially a process that has never been run at the scale being built. Let's start with the basics, while there is a generic term of "start-up", this is actually a sequence of inter-related events that must be completed, as follows:

- Mechanical Completion – this is the phase of the project when the process portion of the plant are constructed and ready to be tested. Very often, secondary items such as office buildings and other non-process portions of the facility may still be under construction.
- Commissioning – first time the equipment is tested for basic functionality. Motors are rotated, valves are stroked, water is pumped around and fermentation systems go through sterile hold testing. Intent is to ensure the system appears ready to operate before live materials are introduced.
- Startup – this is the time when feedstocks are introduced and the system is tested to see if it runs as designed. Initial testing will be basic functionality, working up to increasing operating rate over time. It can typically be weeks or months until a new system is operated at design capacity, even for a short period of time.
- Release to operations – the first three steps are managed by the startup and engineering teams. They are responsible for activities from mechanical completion through startup. Operators should be intimately involved for training purposes, but the responsibility has not been passed to them. Once everything checks out and procedures are in place, there is a formal transfer of the equipment from the startup team to the operations team.

In a traditional project delivery model, each one of these steps would be completed for the entire facility before moving onto the next step. The aggressive timelines that are common in startups do not typically allow this. To save overall project timeline, portions of the plant go through these steps in a rolling manner as they are constructed. Most of the major facilities I have led the startup on in recent years have had the front end of the facility operating before the back end of the facility is complete. This can be done, but requires a strong engineering and process development team, combined with a commitment to work through issues as they arise.

There are many lessons I have learned in facility startup, but the three most common are:

- Run the plant like it was designed – very often, the sheer size and cost of the news facility can get the startup crew thinking timidly and feeling maybe they should try operating the plant in some type of reduced mode. Very often starting in a batch mode, even though the facility was designed to be continuous. This is a HUGE mistake. Running the plant outside of design is very difficult and generates more issues. The plant was designed a certain way for a reason, so run it like it was designed and when issues come up (which they will), work through them. It's the only way to get the plant online.

- Continuous processes do not run well at very low rates – the concept of turndown is important when starting up a new plant. While you want to take things methodically, most processes do not turn down more than about 3 to 1. So, if you try to operate equipment at less than 33% of capacity, it likely will not operate right. Like item #1, you need to be confident in your plan and team, and plow forward. I have seen more issues generated by trying to operate at very low rates than solved.
- It is not a question of if you have problems, it is how many and how you resolve them. Even with significant pilot testing, you will have a plethora of issues arise during startup. Instruments will be determined to be the wrong application, valves will not work, lines will plug and a long list of items will be generated. While frustrating, it is the reality of startup and be prepared for it. It is very common to start up, run for a period of time to generate a list of required modifications, then shut down and make them. If it is being done correctly, each time you restart, you will see improvement and a longer period of sustained operation.

The summary lesson learned is that there is no substitute for a technically strong and experienced startup team. This should be a mix of scientific subject matter experts who know the process and chemistry, but are not necessarily experienced in plant operations. They are then teamed with an engineering team, experienced in installing and operating major equipment. The final requirement is a steadfast determination and resolve to make it work. This type of startup has historically been the most challenging and rewarding moments in my career. I have often told startup teams that it is a monumental task ahead of us, but I guarantee they will look back on it fondly at some point. I then usually clarify that point usually comes 6-12 months after the startup is complete and the exhaustion has subsided.

10. Plant Operations - *It's not grad school*

I began this lessons learned chapter talking about cultural team changes in the early phases that are necessary for the venture to succeed and want to return to the subject to discuss the change from discovery and innovation to ongoing operations. I have had many CEOs over the years passionately tell me that once the production facility is operational that company cannot quit discovering and innovating, and while I agree, I point out that the proper place for that discovery and innovation to occur is in the lab and pilot plant, not by making continual changes to the commercial-scale operating facility. One concept I have learned as I have moved between R&D and operations multiple times is that what makes you successful in one, can often be a problem in the other. Let me explain.

What makes a good scientist and researcher is a drive for discovery and unwillingness to accept any constraints, other than laws of thermodynamics. This is often the DNA make up that allows brilliant people to discover concepts missed by others. Most researchers are motivated by the challenge and thrive in a rapidly changing landscape.

By contrast, what makes a good production process is predictability and reliability. The knowledge that you have a very high probability of success in making your product on time and at the cost you projected. This comes from structure and standard procedures, where most good operations staff thrive.

If these two worlds seem different, they are. Both are valuable and critical for long term success in their own way, but when the worlds collide, it can make for big problems. Basic lesson learned, innovation should be focused in the lab and pilot, and when ready, transferred under a formal management of change process. Many of the development team will be used to doing "on the fly" tweaks to the process at pilot-scale and this can be very dangerous and costly at commercial-scale. I have learned that process development tried for the first time at commercial-scale can result in millions of dollars of sunk costs. This is one of those lessons I have also learned a few times and sincerely hope the warning will be heeded and not proven yourself.

SCALE-UP

What Makes Scale-Up Of Industrial Biotechnology So Difficult?

Two questions often emerge from reading the Lessons Learned on commercializing industrial biotechnology, _what makes scale-up of industrial biotechnology so difficult_ and of course _how can risk be reduced in the process?_ This chapter will expand on the concepts to address these questions.

In Lessons Learned, the third lesson, **_"there is no substitute for a fully integrated pilot process,"_** is often a focus of debate. Some pioneers in the advanced biotechnology industry gained their experience in chemical or petrochemical industries, often rooted in process modeling as its primary scale-up tool. This is significantly different than advanced biotechnology, where scale-up is based on extended pilot operations. The cost and timeline of building an integrated pilot, or demonstration scale plant, challenges new ventures attempting to bypass an integrated pilot, which can end badly. Let me focus on what makes biotechnology processes unique and why pilot testing is so critical to develop a successful scale-up, and thus, commercialization.

First, let's understand how traditional chemical processes are scaled-up by modeling as a comparison. As a chemical process engineer who spent the first portion of his career in the chemical industry, I have been faced with many of the traditional chemical scale-up challenges. Processes such as synthesis of an organic compound and subsequent refinement from a mixture of solvents, where compounds had well documented chemical and physical properties. If there were chemical reactions, it was typically between a limited number of compounds with well-known reaction kinetics and a short list of competing side reactions. This can be modeled very accurately by process simulation software like Aspen or CHEMCAD. Modeling did not completely replace the need for piloting, but often limited the scope of pilot testing to verification of key parameters. This history of success in using modeling processes and then verifying a few separate operating conditions with piloting, gave confidence in this approach.

Now, let's compare that to industrial fermentation-based processes. Industrial fermentation typically starts with feedstocks that are less pure and more complicated from a reaction standpoint than a traditional chemical reaction. Anyone who has seen the massive wall posters of metabolic pathways in very small font knows what I am talking about. It is generally not practical to model the entirety of the individual reactions (and competing side reactions), but rather it's practical to generate an average rate equation for the overall process. While this can be used to represent the process from a "macro" perspective, it will not accurately predict the minor constituents in the fermentation broth that can impact both the fermentation and recovery productivity. This example is specific to fermentation, but the principle equally applies to other bio-based processes.

Given this inability to accurately model biotechnology processes, pilot and demonstration plant operation is the only reliable method to generate the information needed to scale and design equipment. Accordingly, an integrated pilot operation is critical to project success. Here are a handful of my lesson's learned specific to scale up of biologic processes:

Understand your feedstock - If you are planning to use standard industrial sugars like liquid dextrose at commercial scale, you need to make sure your lab and pilot testing accurately represents the feedstock. As an example, if you are buying bags of commercial dextrose crystals to run in your pilot plant, you might be surprised to find out that crystal sugar has a much higher purity (>99.5%) than standard liquid dextrose. Typical liquid dextrose is only 95% dextrose and has 3-5% of other (often unfermentable) sugars including maltose and higher saccharides. These can cause operational issues both in fermentation and downstream recovery. Failing to use representative feedstock during scale-up testing can set you up for big problems later.

The same issues arise when doing fermentation with syngas or digester gas as a feedstock. Often, the commercial business model will require producing syngas by gasification of biomass or municipal solid waste (MSW), yet the lab or pilot will operate on syngas generated from natural gas for convenience. Just like the sugar example above, if the feedstock used in the pilot does not represent the reality of commercial scale, there will likely be operations issues that arise.

The liberal media – A reference to the fermentation media, of course. This is a mixture of trace minerals and vitamins added with water and inoculum at the beginning of the fermentation. Just like humans, most organisms need some trace level of these to support metabolic activity. The hard part is determining how much is required and not just liberally adding it to make sure there is plenty. Optimization of media is not generally a priority at lab or pilot scale, but can become a significant cost and supply constraint in a commercial operation. Just like the dextrose example, commercial supplies of the vitamins and minerals are less pure and can bring contaminants and other compounds that can negatively impact the process.

The "other" problem - as discussed above, it is not practical in a commercial biotechnology process to predict all compounds generated during the fermentation, or that come along with the feedstock. Typical chemical analysis used in engineering scale-up will identify key compounds, but then everything else that cannot be identified gets lumped into a category of "other", often referred to during the design process as "OS" or "other stuff". It is important to note that these compounds are not inert and usually impact the process. The biggest issue usually comes from the unfermentable sugars and the co-compounds that are generated from side reactions. This is one more case where the only way to determine the impact of these compounds is to run the pilot process.

Don't push the rope – hopefully we all learned at an early age that you cannot push a rope, you need to pull with it. The same principle applies to process scale-up. You need to first determine what your commercial scale facility will look like conceptually and use the pilot operation to prove out key parameters needed to build the process (i.e., "pull" the information needed from the pilot). This involves identifying commercial scale equipment that can perform the unit operations you need and utilizing the pilot to generate the data needed to select and design commercial equipment. Trying to just replicate what you have on a pilot scale, without consideration of what is practical at commercial scale (pushing data forward), will not usually result in a viable process.

Determining your key parameters for scale up – It is critical early in the pilot process to determine what information you will need to design your commercial facility and how to generate what you need. Many times, it's proving out whether a factor will impact your process. Think of scaling-up from a standard 300-liter packaged fermenter to a 300,000-liter air lift fermenter. The packaged fermenter is about 5 feet tall and the airlift fermenter could be nearly 100 feet tall. Consider these questions:

- The pressure at the bottom of the airlift fermenter where the air or syngas is dispersed will be much higher than the packaged fermenter. Will the pressure impact how the gas mixtures are absorbed, accessed by the organism and/or the overall fermenter performance?
- Packaged fermenters utilize mechanical mixing while airlift fermenters use the rise of the gas bubbles to induce mixing. How do you determine if the airlift mixing will be adequate?

These are just a few of the items that can only be determined by knowing what you need to prove for commercial scale and how to get the pilot or demonstration operation to generate the information.

Strategic Approach To Scale-Up

The most common question I receive from companies working to scale-up advanced biotechnology is whether their process can operate at commercial scale. My common response is that it is not a question of whether it will work at commercial scale, but whether will work economically. Advanced biotechnology has developed many impressive technologies and the focus of process scale-up needs to involve early stage strategic planning to concentrate on technologies that have a high probability of providing an economically viable process.

In the previous section ***"what makes scale-up of industrial biotechnology so difficult"***, I introduced the concept of <u>*not pushing the rope.*</u> This notion involves identifying commercial scale equipment that can perform the process required and utilizing pilot testing to generate the data needed to select and design commercial equipment. The first step is to develop a concept for the commercial scale facility and use the pilot operation to prove the key parameters needed to build the process (e.g., "pull" the information needed from the pilot). Trying to simply replicate what has worked at pilot scale, without consideration of what is practical at commercial scale (pushing data forward), will not usually result in a viable process.

The dollar test – the first test to use in evaluating process technologies for biotechnology scale-up is called the dollar test. Does the target product sell for dollars per gallon (commodity fuels and chemicals), dollars per pound (specialty chemicals or food proteins) or dollars per ounce (niche chemicals or pharmaceuticals.) The value of the product will dictate which technologies have the best chance of making an economically viable process. For high value items like pharmaceuticals, the entire process toolkit is typically available. However, when making commodity chemicals or fuels that sell for a few dollars per gallon (or fraction of a dollar per pound), it is unlikely that many of the higher end process technologies will be viable. It should not be assumed that success of any specific technology is not possible, but it is worth pointing out that certain technologies are improbable to succeed and require a long development timeline.

Look for footprints along the journey – when hiking in the snow, there is often a choice presented to either use existing footprints that ease the journey or forge one's own path. Technology development is ironically similar, and the determining factor is the same as hiking – is your destination the same as those that have gone before you or are you breaking new ground? This decision needs to be made for each process step. Just because you have an innovative organism or fermentation technology, does not drive a need re-invent the wheel on downstream recovery if an economically viable pathway exists. However, if the destination is different, forging a new path is a given and will need to be resourced appropriately.

Placing your bet – scaling-up advanced biotechnology, by definition, is a valley of death activity. This is the chart that every entrepreneur has seen a thousand times that outlines the negative cash flow that needs to be overcome to reach commercial operations and profitability. Utilization of a strategic approach to scale-up is the best way to mitigate risks and minimize cash flow. When a tentative process is selected for R&D, pilot and demonstration scale development, these activities will become the largest costs during technology commercialization. Selecting which path to move forward on is no different than placing a bet in Las Vegas, the best chance of a successful result is to understand the odds and corresponding risks, to make a decision that provides more reward than risk.

There is always a chart – The following page is a chart that summarizes many of the common unit operations within advanced biotechnology and ranks them for applicability to a range of target products, from fuels to pharmaceuticals. Notice that there are not hard definitions such as "will not work" or "proven", as each technology application is different and thus does not warrant hard lines. Instead, the technologies are rated against the specific products and determined either to be:

Probable – has a history of being used to produce products of this type and sales price category

Possible – technology likely to work to produce required separations, but may be challenged to be economically viable.

Unlikely – even if the technology works at small scale, the lack of availability of corresponding commercial scale equipment makes success unlikely to be achieved.

The fact that a technology has been used successfully in similar commercial applications is not a guarantee of success, but it is certainly a good place to start. Every investment advertisement starts with the disclaimer "past performance is not a guarantee of future success" and while I agree there is no guarantee, past performance is still one of the best predictors of success.

Category	Technology	Fuels	Commodity Chemicals	Specialty Chemicals	Food Proteins	Niche Chemicals	Pharma
Reactors	Anaerobic Digestion	?	✔	✔	✔	✔	✔
	Anaerobic Fermentation	?	?	✔	✔	✔	✔
	Aseptic Fermentation	?	?	?	?	✔	✔
Separation	Clarification	?	✔	✔	✔	✔	✔
	Separator	?	?	✔	✔	✔	✔
	Disc Stack Centrifuge	?	?	✔	✔	✔	✔
	Super G Centrifuge	✗	✗	✗	?	?	✔
Disruption	Acid/Base Treatment	?	?	✔	✔	✔	✔
	Enzymes	?	?	✔	✔	✔	✔
	Bead Milling	?	?	✔	✔	✔	✔
	Homogenization	✗	?	?	?	✔	✔
Sterility	Pasteurization	?	✔	✔	✔	✔	✔
	Sterile Filtration	✗	?	?	?	✔	✔
	Irradiation	✗	?	?	?	✔	✔
Purification	Solvent Extraction	?	?	✔	✔	✔	✔
	Distillation	?	?	✔	✔	✔	✔
	Crystallization	?	?	✔	✔	✔	✔
	Microfiltration	?	?	✔	✔	✔	✔
	Tangential Flow Filtration	✗	?	?	?	✔	✔
	Chromatography	✗	?	?	?	✔	✔
Drying	Spray Drying	?	?	✔	✔	✔	✔
	Fluidized Bed Drying	?	?	?	?	✔	✔
	Freeze Drying	✗	?	?	?	✔	✔

Increasing Product Value →

Likelihood of Technology Succeeding

Probable	✔
Possible	?
Unlikely	✗

Fermentation Development

Fermentation is the core technology behind most industrial biotechnology, using a microbe to convert a carbon source (typically sugar) into a product without animals or crude oil. The first important concept in commercializing a new fermentation-based biotechnology is that there is no substitute for a fully integrated pilot process. While many chemical processes can be modeled, fermentation relies significantly on pilot testing. This section will focus only on fermentation, but keep in mind that it will need to be integrated with downstream recovery to have a successful scale-up, which we'll discuss later. This section will provide a high-level overview of fermentation scale-up, followed by a deeper dive into the design specifics.

The second concept that needs to be understood in the early stages of fermentation scale-up is **fit for purpose.** There are many relevant fermentation design standards, and each has its benefits and downsides, what is important is to understand which is appropriate for your process.

- Pharmaceuticals are the highest standard. Fermentations are expected to be run in a manner that produces a "clean batch", meaning there are no foreign organisms at the end of the fermentation, other than your target organism. This is the highest capital cost and hard to replicate at commercial scale for industrial fuels and chemicals.
- Industrial biotechnology uses pharma as a guide, but makes risk-based decisions to reduce the cost and make it easier to scale-up. There is always a hope, but not usually an expectation that fermentation batches will be clean. Many products have limits on allowable non-target organisms that can be present.
- Food applications are a significant focus area in advanced biotech today. It is important to understand how food production is different than traditional biotechnology. The focus is less on eliminating any contamination, but rather a focus on controlling both total foreign organism count, and ensuring certain restricted organism (salmonella, ecoli, listeria, etc.) are not present.
- Traditional ethanol is the low-cost fermentation standard and very different in design from most advanced biotechnology. Sterile design is less of a concern and there is more of a reliance on the core organism out-competing other organisms.

Let's looks at a more specific example of what needs to be determined and proven during scale-up of fermentation. At the lab scale, you can buy a fermenter as a package. At commercial scale, the system will need to be designed from the ground up and key criteria need to be determined before design can be completed.

The first decision is usually the type of fermenter, whether you will be using an agitated fermenter, or whether an airlift or bubble column will fit your process. Oxygen transfer rate (OTR) will be a key factor in determining the type of fermenter. High OTR processes are more likely to require mechanical agitation, while lower OTR processes can use airlift or bubble column, which generally have a lower capital cost. Cooling is very important. Whether it will be done by internal coils, external jacket or external heat exchanger. Each has cost and sterility implications, and sterile design can be one of the hardest to quantify and is based on how prone your process is to contamination. Processes that are favorable to foreign growth (ie. neutral pH, sugar feed and slow doubling times) will need a higher standard while processes that are less favorable (high or low pH, non-sugar feed and fast doubling times) require less. Finally, how you will clean and sterilize the system, by steam or chemicals, along with a strategy to control the organism is you are using a GMO is necessary.

The final key factor in fermentation scale-up is to develop a concept for a commercial facility and pull data from the pilot, not try to replicate the pilot system at larger scale. Feedstock is a common example. The type of feedstock and purity has a direct impact on the fermentation and recovery strategy, thus using representative feedstock early in the process is critical. I have seen many cases where the pilot is operated on re-hydrated (bagged) dextrose, which is over 99% pure, while the commercial facility is expected to operate on liquid 95 DE, which has significant unfermentable sugars. This can often result in a process that operates as desired at the pilot scale, but faces significant quality issues at commercial scale.

It is important to understand what success looks at the beginning of the scale-up process. The primary consideration is product specification that will be met as the process is developed. It is equivalent of knowing your destination before you start a trip. If you do not know where you're headed, finding your destination will be very difficult.

As a general guide, OTR required by the microbe will usually dictate the style of fermenter. Aseptic fermenters are a pressure vessel, while an ethanol fermenter is designed more like a storage tank. Because of this, the maximum size for an aseptic fermenter is much smaller than ethanol. It is important to note aseptic fermenters are conventionally listed in liters, different from corn ethanol which is in gallons. These conventional references in industrial publications can create an issue for the casual reader, if the differences are not noted. Converting the aseptic fermenter from liters to gallons, it's approximately 130,000 gallons or only about 15% of the size of an ethanol fermenter. One of the biggest surprises for many who started doing work across the range of types of fermentation is to realize not only that corn ethanol fermentation is anaerobic (without air), but how much air goes into aseptic fermentation, and the capital and operating costs it represents.

Fermentation temperature can often drive a requirement for chilled water. This can be both a significant capital cost and an on-going operating cost issue. Cooling for aseptic fermentation can be done internally or externally, while ethanol is external.

Finally, there is large difference on equivalent capital cost. A budgetary installed cost number for a commercial scale aseptic fermenter is in the range of six to ten million dollars and a corn ethanol fermenter around five to seven million. However, when you consider how much larger a corn ethanol fermenter is, the cost per unit volume is roughly ten times higher for aseptic than corn ethanol. The focus of most advanced biotechnology ventures is to figure out the most cost-effective strategy in a sea of possibilities, to commercialize its technology.

Making the tough decisions when scaling-up fermentation involves defining key operational criteria and prioritizing in order of criticality. A sample of key fermenter design criteria is below, which compares an aseptic fermenter to corn ethanol.

Criteria	Aseptic Fermentation	Corn Ethanol
Fermenter Pressure	30 – 50 PSI	Atmospheric
Typical Commercial Size	500,000 liters	1,000,000 gallons
Cooling Configuration	Internal Coils/External Exch.	External Heat Exchanger
Air Requirement	Sterile Air	None (anaerobic)
Mixing	High Shear Agitation	Low Shear Mixing
Surface Finish	Mid to High Polish	Standard Mill Finish
Sterile Boundary	Robust, Heat and Filtration	Limited
Cleaning/Sterility	Clean then Steam in Place	Clean in Place
Typical System Cost	$6M - $10M	$5M - $7M
Typical Cost per Liter	$16	$1.6

Fermenter Style Options

Selecting the right style of aseptic fermenter is critical and worth discussing in more detail, specifically the two most likely fermentation configurations, which are agitated and bubble column. As you can see from the figures, the agitated fermenter has a mechanical mixing system and is short and wide. By contrast, the mixing in a bubble column comes from the rising air bubbles and the inside of the fermenter is open. In the case of a bubble column, they are typically very tall are slender. There are advantages for both systems.

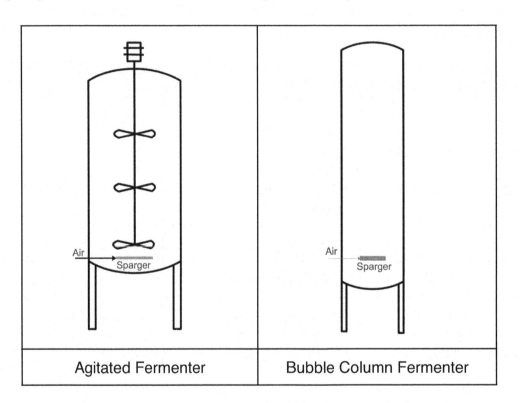

| Agitated Fermenter | Bubble Column Fermenter |

Typically, agitated fermenters are used for higher oxygen transfer fermentations, with liquid feedstocks like sugar or glycerin. By contrast, bubble columns are used in lower oxygen transfer fermentations and a good fit for gaseous feedstock like syngas or methane. On a cost basis, agitated fermenters are usually more expensive and have a smaller maximum size, so if an airlift fermenter will work in your application, it is likely the most cost-effective option.

The ability to fabricate fermenters offsite for a new plant can also be an issue for commercial facilities. It is usually more cost effective and higher quality control if you can fabricate the fermenters in a shop and transport them to the site. As you get to commercial scale fermenter size, this is more difficult. Often the short and stubby aspect ratio of agitated fermenters can make them hard to transport, while bubble columns are longer and narrower, generally making them easier to transport. Not a deal breaker, but definitely a consideration when evaluating options. A detailed summary of the pros and cons for each style of fermenter follows:

Agitated	Parameter	Bubble Column
~3:1	Aspect Ratio (height/width)	~7:1
Mechanical Agitation	Mixing Method	Bubble lift or pump around
• Higher OTR and/or viscosity • Liquid Carbon Sources • High Cell Concentrations	Best Fit	• Lower OTR and/or Viscosity • Gas-based Feedstocks • Lower Cell Concentrations
• Higher Capital Cost • Smaller Max Size • Onsite Fabrication	Sterile Air	• Lower Capital Costs • Larger Fermenters • Offsite Fabrication

Agitation is a major design consideration. Mixing is key to oxygen transfer, but it's important to consider that viscosity, cell density, pressure and heat removal are also important. Heat removal is often seen as a limiting factor. It is often stated as an oxygen transfer limitation, while the actual issue is inability to remove heat. When looking at options, the typical benchmark is mixing power per unit fermentation volume. This provides an apple to apple comparison. When looking at an agitated fermenter, a primary consideration beyond horsepower is configuration of the agitation blades. How many blades, where are they are located, and whether the blades are propellers (known as pumpers) or Rushton, primarily for gas/liquid mixing.

People new to aerobic fermentation can be surprised by the amount of electrical power required. I can tell you this was my experience. On the first biotechnology project I developed, I was convinced the power requirement was a typo. I came from the world of chemicals and biodiesel, where a commercial scale facility may use 2 or 3 megawatts of power. I can tell you it is not uncommon for a commercial scale fermentation facility to use 20 or 30 megawatts of power, almost an order of magnitude higher. The power hog is generally fermentation, with chillers and air compressors as a major contributor. I had used both chillers and compressors many times before, but not at the number and scale required in fermentation.

Safety is king. Both syngas and methane are common feedstocks for fermentation, but they present a different set of challenges and advantages than typical aseptic fermentation on sugar. The first key factor is that the gasses are flammable, so they present a safety issue. This safety risk amplifies if the fermentation is aerobic, as there is a much higher potential for an explosive mixture. One of the benefits of methane and syngas fermentation is that there are many less organisms that can compete in the matrix, so sterility is much less of a concern. This allows for these fermenters to often be designed more like traditional ethanol than aseptic. Heat removal can be much higher for gases, especially methane. It is not unusual for a methane fermentation to have 5-10 times the heat generated as other fermentations. Finally, syngas consistency is important to many fermentations and can be an issue depending on the source. Most organisms require a consistent hydrogen to carbon monoxide mixture.

Sterile Design Basics

If you've spent any time in commercial fermentation facilities, then you already know it's a hot bed of contaminants and has steam coming out all over the place. This is because heat (in the form of steam) is just one of the keys to keeping fermentation sterile. As companies look to scale industrial biotechnology based on aseptic fermentation, many look for ways to cut capital costs and the robustness of sterile design is often a target. Beware, removing items such as steam sterilization can cause you to get burned in a different manner (financially). It often comes down to spend now or pay later, let's explore.

Easiest way to determine the potential impact of aggressive decisions on sterile design is to take the current version of the technoeconomic model and change the "failed batch" rate from a typical 1 - 5% failed batch rate for a mature operation to 30% and see how the economics of the process change. All the labor and raw materials will be expended, but 30% or more of the production will not be usable. Even worse, the facility will likely have to pay increased treatment or disposal costs to get rid of the bad batch, while slowing down overall plant production. This is the downside of a poor sterile design and the cold hard reality of being overly aggressive.

For aseptic fermentation, where there is an expectation of only one desired target organism at the end of a batch, sterile design is best summarized by creating a boundary that keeps all non-target (foreign) organisms out. This is done by ensuring the fermenter is clean and sterile, with everything coming in or going out being sterilized by one of the two primary methods, heat and filtration. Heat above 130C is used to kill all organisms. This can be done by sparging steam into the fermenter, or in the case of streams like sugar, heating them up for a short period of time to target temperature. For items that are heat sensitive, they are sent through a 0.2 micron filter (commonly known as "sterile filter" in the fermentation industry), that ensures no organism can make it into the system. The following graphic depicts how a sterile boundary is maintained.

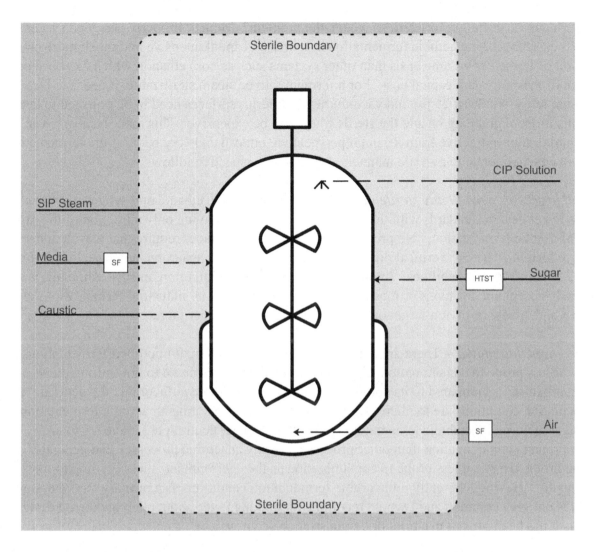

Sterile Filters (SF) are 0.2 micron filters that provide a sterile barrier for gasses/ liquids entering the system

High Temperature Short Time (HTST) heat sterilization system thermally kills organisms with 130C heat

Clean in Place (CIP) is a hot (60C) caustic solution sprayed in the top to clean the fermenter to bare metal

Steam In Place (SIP) is steam directly injected into the fermenter to heat to 130C

Surfaces Matter - fermenter and piping surfaces are polished to keep organisms from grabbing on

Double block and bleed is a system where all valves in the system that are open to atmosphere are double-valved with a steam bleeding through a small drain so no organism can make it into the boundary

This sounds very straight forward, so what's the problem? Money: Properly designed systems are expensive. Aseptic aerobic fermentation systems are capital intensive and much more on a per unit of fermenter volume basis than other systems such as corn ethanol, which can be about 1/10th of the equivalent capital cost. For a fermenter to be steam sterilizable, it needs to be pressure rated to at least 25 psi and vacuum rated. Interior surfaces need to be polished to keep organisms from grabbing on and the sterile filters can be expensive. This makes using lower standard designs attractive from a capital perspective, but can end very badly if the fermentation system cannot operate in a sterile manner. Additional perspective follows.

Do all processes need highly sterile design? – No, systems like ethanol and syngas operate at much lower design standards with their fermentation systems having only a fraction of the capital cost of aseptic fermentation. No pressure vessels or metal surfaces require that you should see your reflection. There are critical differences though: ethanol is anaerobic and has ethanol that limits organism growth abilities. Another factor is presence of other organisms: while this is not typically acceptable for foods or most aseptic fermentations, many industrial fermentations can have a high presence of non-target organisms and still meet quality targets.

Differences that matter – There are many factors that make different microbial fermentations more or less prone to contamination, but at a high level, it comes down to how strong the target organism is compared to its competition, and how favorable/unfavorable the general fermentation conditions are to microbial growth. Processes operating an aerobic fermentation on sugar at normal metabolic temperatures (25-40C) and near neutral pH (above ~5-6) are more prone to contamination than other processes. Many anaerobic processes that generate acidic broths are much less prone to contamination as the environment is less inviting to most organisms. Having inherently unfavorable fermentation conditions as a primary sterile design basis is not very common and I would recommend thinking twice about alternate sterile designs unless it has been proven in a pilot or demonstration scale system.

A fork in the road – The hardest part of sterile design is that is must be decided up front. A fermenter that is purchased without the ability to be steam sterilized cannot practically be retrofitted. If the system does not run sterile, there are limited options to correct it.

Capital cost savings from a decreased sterile design are real and I do not mean to downplay the value they provide to an early stage company, but simply offer a framework that should be considered before heading down what can be a risky path.

Downstream Recovery

Advanced biotechnology was born out of advances in fermentation and synthetic biology, often making fermentation the technical focus of many companies. However, as the overall process is developed to make a product, the recovery of the product from fermentation broth is often more complicated and costly than fermentation. Unfortunately, downstream recovery does not often receive similar development resources at early stage companies and can cause issues later in the process.

In developing a technical approach to recovering the product from fermentation broth, there are two major pathways; products that are secreted into the broth and products that are inside the cell at the end of the fermentation. Both of these systems will use similar unit operations, but they are configured differently. As you start to develop a recovery process, overall yield and product quality need to be the primary focus. These will need to be optimized to end up with a process that is commercially viable. As noted on the figure below, there are two major pathways to purification, depending on whether the product is secreted or inside the cell. These are shown on the following block flow, with the difference being addition of a cell disruption unit operation that is required for inter-cellular products.

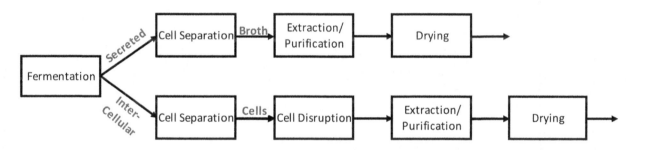

The table below shows the typical unit operation used in advanced biotechnology. The remainder of the presentation will do a deeper dive into the strategic considerations in determining which technology works best for a specific process.

Separation	Cell Disruption	Extract/Purify	Evap/Drying
Disk Stack Centrifuge	Bead Milling	Solvent Extraction	Spray Drying
Decanting Centrifuge	Heat/Steam Explosion	Distillation	Double Drum
Filtration	Acid/Base Treatment	MF / UF/ TFF	Rotary Steam Tube
	Enzymes	Chromatography	Flash Evaporation
	Homogenization	Crystallization	Multi-Effect Evap

The most universal operation used in downstream recovery is concentration or separation by centrifuge. It is important to understand the types of centrifuges and how they scale, as they are usually more complicated and less flexible than first thought. The primary types used are disc-stack, decanter or high-G tubular. The size of the cells being separated is a key factor in determining the appropriate equipment. For most advanced biotechnology, the cells are generally in the .5 to 5 micron range and typical fermentation broths are 2-10% dry cell weight concentration. The chart below provides a visual representation of the typical operating ranges for the centrifuges discussed.

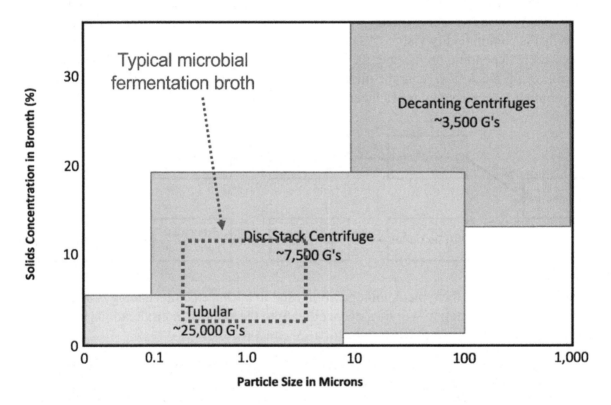

Let's look at lessons learned scaling centrifuges. Particle size, G force and the amount of total solids will be the key criteria in determining the right centrifuge. I often see pilot plants using tubular high G centrifuges that provide good separation, however they are not scalable and do not have a commercial scale equivalent. This can make using them during scale up a mistake. Decanters are desirable as they have higher throughputs and lower capital cost than disc stacks, however they are not typically a good fit for most advanced biotechnology. Though there are exceptions, they are not usually worth a significant development effort. Disc stack centrifuges are the most widely used, however there are many ways to configure them. When sized correctly for flow and solids handling, they can operate continuously for years. By contrast, when the wrong configuration is selected, it can produce significant mechanical and quality issues.

Moving on to cell disruption, I am going to provide a simplistic example of the options and implications for the overall process. In lieu of cells, let's consider a visually easier example; pistachios. Like our cells, there is something inside we want (the nut) and there are various ways to disrupt the shell to get to the nut. How we disrupt the shell will impact the recovery and quality of the nut. In the case of secretion, the nut and shell are both intact. Bead milling is a process that cracks the outer shell releasing the internal material. Enzymatic hydrolysis is similar, but done chemically, not mechanically. Acid and solvents can be used to disrupt the cell, but perhaps far too aggressive. Finally, homogenization is the most aggressive disruption technology and completely deconstructs the cells. As the cell is disrupted to a higher level, the resulting matrix is usually more difficult to recover the target product.

There is a broad range of disruption technologies - determining which is the best fit will depend on the overall process. Organisms that secrete the product into the broth do simplify the recovery process, but often have a lower yield. The higher the level of deconstruction to the cell, the more complicated the recovery process becomes. One secondary item for disruption technology consideration is the impact on spent biomass. Most biotechnology processes generate biomass in the form of the spent cells. This can have some value, however the more the cell is deconstructed, the less valuable the material becomes.

Extraction and purification have the widest range of technology options in downstream recovery. It is not usually a question of whether the product can be recovered, but whether it's economical. The fermentation process is likely not a static operation and if changes are made to improve the fermentation process, those changes will impact the downstream recovery efforts. Filtration systems have made significant advancement in recent years and can be more cost effective than previous versions. That said, it's important during development to understand filter life and ability to be cleaned, as replacement filters are a major cost. Don't be too quick to discount legacy technologies like solvent extraction or distillation. They can often be the most robust and cost-effective solution.

As recovery processes are developed, wastewater must be optimized. Fermentation processes generate significant wastewater and downstream recovery has the potential to add to it. Optimization is critical as wastewater can have significant capital and operating cost impacts. It is common for wastewater treatment to add 10-15% of overall project cost. As the process is developed, it is critical to get representative samples from the pilot operations and analyze them for key criteria. This will be valuable for permitting and facility design.

Evaporation and drying are usually the highest energy use in downstream recovery and a significant capital cost. The best option will depend on whether the material to be dried is water-like or a semi-solid. When looking at various technologies, it is best to compare them on water evaporation rate and removal bases. Evaporation is usually more energy efficient that drying alone and is often utilized in a 2-stage combined drying process. Also, direct fired drying systems will be more energy efficient than indirectly heated systems with steam. When considering drying options, it's important to understand which energy sources (natural gas, high pressure steam, etc.) are available at the site, as this has a trailing impact on equipment selection.

Finally, moisture specification. Most solids with less than 10% moisture appear dry, but if specifications require lower levels of moisture, this can be hard to achieve. The last few percent of dryness can often be very hard to achieve. Skin temperature of various drying systems can impact color and quality and should be considered. On food products, direct fired systems can impart flavor issues and need to be considered. Solids handling, once dried, can be difficult and needs to be a focus. Freeze drying, often used during pilot operations to meet the real need to generate samples, is not usually economical at commercial scale.

To summarize, there are a handful of common pitfalls that should be avoided when scaling up downstream recover. High G centrifuges and freeze dryers are not scalable and when used at pilot scale, can misguide long term decisions. On cell disruption, while homogenization is often used in pharmaceutical processes, it is not typically economically viable for commodity products. Similarly, tangential flow filtration (TFF) and chromatography can have issues being economically viable, however advancements have closed much of the gap. Keep a focus on wastewater throughout the process given its impact on capital and operating costs.

Evaluating Commercial Readiness

In the previous section on what makes the scale-up of industrial biotechnology so difficult, the need for a fully integrated pilot plant was developed and common pitfalls on the road to commercialization were explored. Now that we've made some decisions, we need to answer the obvious question: "Are we ready to spend big money on commercializing our technology?"

Here, we'll provide a framework to allow ventures to assess their current state of process development and readiness. While product sales and other very real commercial needs drive technical requirements; here we'll focus just on the technology portion of the evaluation.

The evaluation process begins by determining expectations. Ventures come prepared to present a large data set of fermentation titers, unit operation yields and throughputs, along with product quality information, asking if they are ready to take the next step in commercialization. While technical data is important to the evaluation process, the first question to be answered is what does success look like? A company cannot determine if they are ready to start the next step in the commercialization journey unless they understand the realistic goal. The more rigid and aspirational the expectation, the longer it will take and the less likely the venture is to meet the goal. Many years commercializing advanced biotechnology has taught me that being aggressive or conservative are both valid commercialization strategies, as long as the approach to commercialization matches the company goals and the expectation of success: let me explain.

The most common factors that come into play when determining the expectations of scaling-up a process are capital cost, project timeline, manufacturing cost and projected production capacity. The level of certainty that is required for each of these criteria, at the front-end of the process, will set the overall definition of success. This can generally be broken into 3 categories:

High probability (slow and easy wins the race) – having a high confidence in the capital cost, project timeline, manufacturing cost and ultimate capacity of the facility before commercialization begins. Common among large, established businesses. There is an expectation that the project cost will be fixed, the timeline highly probable and the facility will produce near capacity soon after being released for operations. This is the typical expectation for debt funded projects like traditional corn ethanol, but is generally aspirational for advanced biotechnologies. As new biotechnologies represent processes that have never before been built, the technology risk makes reaching this standard very difficult.

Balanced (no risk, no reward) – willing to accept some level of uncertainly in capital cost, timeline, manufacturing cost and schedule, usually dependent on risk-based decisions and trying to bound the uncertainty. Company has reasonable equity reserves to be able to take modest risks and can absorb increased capital cost, extended timeline and slower than projected production ramp-up. Project can be done quicker than the high probability approach, if a level of uncertainty is acceptable.

Speed to Market (ready, fire, aim) – ability and willingness to accept higher level of risk in commercialization to keep first mover advantage. Often early leaders in technology areas that are able to raise significant venture funding have such a forward lean towards rapid commercialization, they are willing to accept cost, schedule and production risks to get product on the market as soon as possible, maintaining their first-mover advantage over competitors. This typically involves making the go/no-go decisions on major portions of the facility with imperfect information and filling in the gaps (hopefully) during engineering and construction. Only a viable option for equity funded projects.

As noted previously, each of these approaches has advantages and disadvantages. There is no generic "best" option, but rather the option selected must fit the company business objectives. The problem comes when companies want the high probability outcome without putting the time and resources in up front.

One final point I want to make in preparation for the technology readiness assessment outlined below is what I call the ***driver's license test.*** There will be many questions that deal with current state of technology and what has been proven to date. It is critical that realistic data be used in the assessment and to highlight the concern, let's consider a person's weight on their driver's license as an example of perception versus reality. Most people (myself included) have a weight on their driver's license that represents what they believe they should weigh or what they aspire to. Unfortunately, that usually conflicts with the blinking digital numbers between their toes when stepping on a scale. In conducting a readiness evaluation, use the numbers on the scale, not the driver's license.

Now, let's move into the details of evaluating the state of technology and readiness for commercialization. As an example for demonstration purposes, let's use the di-methyl glop (DMG) process I have noted earlier in this handbook. DMG is an industrial chemical made by fermentation, then separated by centrifugation, purified by crystallization, dried and packaged. We will assume the process is currently operated at a 45,000 pound per year pilot plant and the goal is to build a 2,000,000 pound per year commercial scale facility. A block flow of the process is shown below.

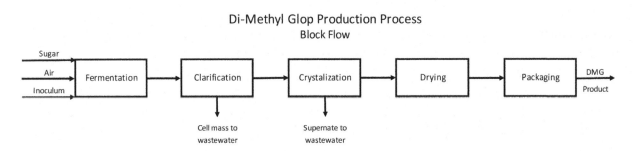

In looking at any advanced biotechnology, there are 5 primary criteria that are used to evaluate readiness for process commercialization as outlined below. None of these criteria individually are a deciding factor, but rather in aggregate are the information needed for the analysis and will be the inputs required for the evaluation tool:

Number of unit operations – The top criteria that usually determines the level of complexity in process development is the total number of unit operations (processing steps) involved in the overall production process. The more unit operations, the higher the level of effort required to make a fully integrated process that functions effectively with them. In the DMG example, there are five unit operations, in the lower range for most advanced biotechnologies. This is primarily because the DMG process has an organism that excretes the final product. In many processes, the product is inter-cellular and requires the cells to be disrupted (homogenized, bead milled, acid/base, etc.) and then isolated (MF, UF, TFF, chromatography, etc.). This often expands the total unit operations in the overall process to ten or more.

Scale-up factor – This is the growth in size between where the process is currently operating and the proposed production level. In the case of DMG, the commercial scale facility is proposed for 2,000,000 pounds per year and the pilot plant that has been operating has an annual production rate of 45,000 pounds, so dividing the two give a scale-up factor of 44:1. The larger the scale up factor, the more risk involved. It is also possible to look at the scale-up factor of each unit operation and then average them. This approach can show a lower level of risk where scaling up involves multiple pieces of identical equipment (fermenters, centrifuges, etc.)

Status against key process criteria – Every process has a handful of key criteria that are the most important technical milestones needed for the process to be commercially viable. These are typically the items in the techno-economic model that have the most impact on total manufacturing cost when a sensitivity analysis is run. Fermentation titer, product yield per unit feedstock and overall recovery yield of product are some of the most common examples. To determine the overall status against process criteria, divide the current status of each criteria against its target, then average them. For the DMG example, in the case of titer, the target is 50 grams per liter, but the pilot is currently operating at 35 grams per liter, so the ratio is .70 or 70% of target. Example calculation for each criteria and overall averaging is shown below:

	Pilot	Commercial	Ratio
Titer (g DMG/l)	35	50	0.70
Fermentation yield (g DMG/g glucose)	0.25	0.40	0.63
Recovery yield (g DMG recovered /g DMG produced	76%	93%	0.82

Overall Process Key Criteria Average	0.71

The risk in key criteria comes both from how far any individual criteria is from the target and also how far the overall average process criteria is from the target. In cases where there is a significant difference of importance between criteria, the individual criteria can be weighted in the calculation, as appropriate.

Fully integrated versus independent operation – It is critical when gathering data at bench or pilot scale for a process, that the process is fully integrated, meaning it is run in a continuous manner in the same location from feedstock to final product. This is the desired standard. In some cases, the process needs be broken into pieces with some portions of it (say fermentation) happening at one site, with broth shipped to another site for downstream recovery. While the second approach does demonstrate the process can produce product, the data generated is of less value than a fully integrated process.

Length of testing at previous stage – The length of time that a process is operated in a continuous manner is very important to determine the quality of data for minimizing risk in scale-up. As contaminants can build up in a process over time, it is common for some problems not to be demonstrated until after weeks of operations. 1,000 hours of continuous operation of the process at the planned conditions is the industry standard. Processes can (and have) been scaled with less operational time, but it does come with higher risk.

When conducting an assessment of process readiness, it is usually helpful to generate a framework diagram showing the block flow of the process at pilot, demonstration scale and commercial, with corresponding production levels. This allows for a more intuitive visual assessment of the magnitude of scale up and highlights if any of the unit operations have changed at the different stages of development.

DMG PROCESS SCALE-UP FRAMEWORK

PILOT
1,310 lbs/year

34 : 1 scale-up

Fermentation → Clarification → Crystallization → Drying → Packaging

Need 1,000 hours of representative operating data from pilot

DEMONSTRATION
45,000 lbs/year

44 : 1 scale-up

Fermentation → Clarification → Crystallization → Drying → Packaging

Need 1,000 hours of representative operating data from demonstration

COMMERCIAL
2,000,000 lbs/year

Fermentation → Clarification → Crystallization → Drying → Packaging

Commercialization Readiness Scorecard - The excel based scorecard takes the key information discussed in this section and provides feedback on state of readiness. An example of the input variables is shown in the following figure. Given the level of information required, it is a high-level assessment, but can provide key insight to which areas are of highest risk and where resources are best spent to close technical gaps. Electronic version of the commercialization scorecard can be found at www.warneradvisorsllc.com.

COMMERCIALIZATION READINESS SCORECARD
Inputs

Expectation	(pick one and insert corresponding number in box)
High Probability - expect high level of confidence in capital cost, timeline and ultimate capacity of facility. Insert "1"	
Balanced - willing to accept some level of uncertainty in capital cost, capacity and schedule. Insert "2"	
Speed to Market - strong drive to get into larger scale production as soon as possible, willingness to accept significant uncertainly. Insert "3"	

2

(1, 2 or 3)

Number of Unit Operations	(insert number)
Total number of unit operations from feedstock to final product	

5

(3 - 15)

Scale-up Factor	(insert number)
Growth in size between current operating scale and proposed project	

44

(1- 1,000)

Status of Key Criteria	(insert decimal number)
Average ratio of key process criteria when compared to commercial target	

0.70

(0.0 - 1.0)

Degree of Process Integration	(pick one)
Fully Integrated - process is being operated from feedstock to final product in a continuous process, all on the same location. Insert "1"	
Partially integrated - process is conducted with pieces in multiple locations, attempting to mimic a fully integrated process. Insert "2"	
Independent - unit operations are run individually and modeled based on mathematical representation of how they form an overall process. Insert "3"	

2

(1, 2 or 3)

Length of Testing	(hours of continuous operation)
Operational hours of testing at the previous (pilot or demonstration) scale that represents the commerical scale process	

547

(10 - 1,000)

Presenting Biotechnology Scale-up During Fundraising

Before we begin to discuss how to present the maturity of a given technology during fundraising, let's review what we mean. It begins with development of the proposed process at bench scale, works through pilot and/or demonstration scale (depending on the type of technology and risk tolerance), and ends with a commercial-scale facility. When looking at scale-up, it is always best to start at the end and work backwards. This requires having a concept of what the commercial facility will look like. The concept will drive early stage capital cost and operating cost estimates, a high-level roadmap for development. These will identify risks and key cost criteria. Scale-up is a journey, so treat it like a road trip. If you do not have a clear idea of your destination when you start, you will likely run out of gas money before reaching your destination.

How to forecast key criteria for a process during fundraising is a difficult subject. The most common example is summarized with the following graph.

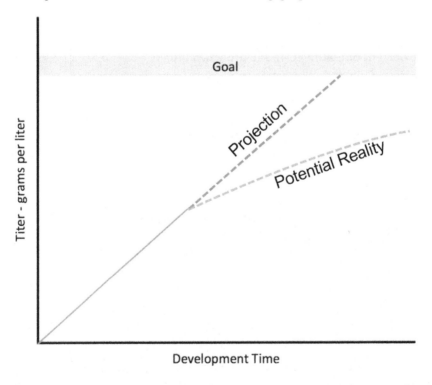

For fermentation-based processes, the ultimate titer at the end of the fermentation is often a key cost factor. The graph shows early success with performance at roughly half the goal. Assuming the level of success continues, a projection is developed. Many years of scaling these technologies has also shown a potential reality of performance leveling off. Even if the goal can be reached, it may be a much longer time frame than expected. During early stage development time is money, so it's important to project milestone timelines, because in reality, it represents cash burn.

The tension between cash burn and timelines can sometimes create what can be politely characterized as "overly optimistic" when describing current state of technology development. In the end, being overly optimistic hurts the company and the overall industry, so I encourage everyone not to get too far over their skis in projections.

So, what is the appropriate level of optimism during fundraising? Let's revisit that driver's license test outlined in a previous section. Most of us, myself included, have a weight on their driver's license that represent what they believe they should weigh, more than what they actually weigh. Think of it what you would weigh if you had an optimum mix of eating and exercise. What it usually is not, is a weight you could never reach, no matter what. Forecasting technology achievement is the same, make sure what you project has a credible plan to reach the stated goal.

A close cousin to cash-burn is assuming away risks. Key risks will usually be the areas of technology with the lowest level of development, or the highest projected cost. Identification allows for risks to be managed and also justifies the funding that will be required to mitigate them. Transparency of risk is appreciated by funding sources as in the end, it's all about financial modeling. The potential funders will build a model of your company and process. Risks will be modeled and if you don't identify them, the funding source will, which is often not the best representation of the potential for success.

Finally, a quick refresher on statistics, as this often comes up in my discussions with funding sources. When we think of a bell curve, the center represents the most probable outcome. As we make more aggressive assumptions in items like capital and operating cost estimates, they should still fall under the bell curve, but may represent the 10th or 20th percentile (thus only 10-20% likely to be reached). This is an example of the aggressive nature of forecasts and is important to not the outcome is _possible_, but not _probable._

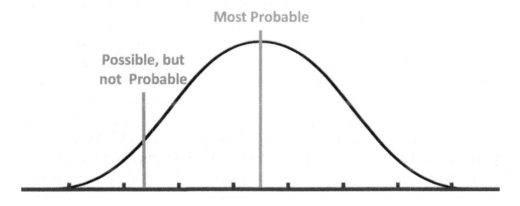

Biotechnology Scale-up Hacks

Not enough time or money to scale-up following the traditional rules of advanced biotechnology (10:1 scale-up factor, 1,000 hours of integrated pilot/demo operation, etc.)? Welcome to the reality of the biotechnology start-up world. The common question in that situation is: *if I do not have all the resources to follow the traditional approach, how are my resources best used to reach commercial success?* Let's explore.

I am compelled to start with the disclaimer that streamlined approaches (what I call hacks, but some would characterize as shortcuts) should be considered by need, not as a primary plan. There are risks involved and reasons why the traditional approach should be the starting point, but reality often requires different approaches. A good place to gain perspective is to look back at the last industrial biotechnology wave and see what lessons can be learned.

The span from 2006 to 2014 was a period of unprecedented growth and expansion in advanced biotechnology, with the last few years being a retreat to focus more on technology development than commercial deployment. The last commercialization wave was fueled significantly by federal stimulus money. While the intentions for ensuring projects were ready for prime-time were sincere, the realities of the high influx of grant money pushed many to build larger plants than the technologies or markets warranted. The large size of biofuel facilities, combined with softening oil prices, produced large-scale facilities that were not able to remain viable. This is a scenario we self-committed biotechies do not want to repeat.

Before outlining what some would characterize as short-cuts, it is important to understand the industry accepted practices on scale-up to provide a baseline from which to start. When making risk decisions during the scale-up process, it is critical to differentiate outcomes that are undesirable from those that are unacceptable. An example of an undesirable outcome is facility capacity, where you end up with a facility that can only produce a fraction of what it is designed for. This is bad, but usually not catastrophic. The economics will shift some, but as the process is refined, there is potential to ultimately reach design capacity. By contrast, an example of an unacceptable outcome is a process that cannot make product that can be sold. This can be for various reasons, but if not overcome, will cause the venture to fail. As we do risk-based decision making, it is important to focus on items that have undesirable downsides rather than unacceptable ones. With that perspective in mind, let's consider the following hacks that have the potential to provide a faster and less costly route to successful commercial operation.

Go big or go home – How large of a scale-up factor from pilot to demo, or demo to commercial, is a common area of focus. While debt financing looks for a maximum scale-up of 10:1 to limit technology risk, many equity-funded biotechnologies have been able to successfully scale at 100:1. The important point when considering options, is that being too aggressive is more likely to result in a facility that does not provide adequate capacity (undesirable) than a facility that does not produce adequate product quality (unacceptable). This makes scale-up factor a target area if higher risk is acceptable.

How much is enough – Talking about data of course. A critical part of scale-up is not only operating an integrated pilot or demonstration process, but generating enough data to de-risk the scale-up process. The industry standard is for 1,000 hours of continuous operation, or roughly 40 days. The risk involved in reducing operating time is a hybrid of undesirable and unacceptable. The unacceptable risk comes from the period of operation not representing an equilibrium case, which is process specific. Some processes come to equilibrium (including recycle streams) in days, while complex processes can take weeks to reach an equilibrium state. This will determine the level or risk. The undesirable risk comes in the form of capacity of the scaled-up process.

Full integration – Critical during scale-up, is for the entire process to be accurately tested, from feedstock through final product, with recycle streams included. This is the only way to get an accurate representation of what the process will operate like and the product quality standards. Modeling is a great tool, but I can tell you from personal experience, just because a process runs well in a simulation model, does not mean it will run the same in commercial scale equipment. Failure to demonstrate full integration can result in a process that fails to produce a product to specification, an unacceptable outcome. While early stage scale-up may involve processing at multiple sites, care needs to be taken to ensure it accurately represents the integrated process.

Making risk-based decisions to optimize scale-up efforts is often the reality that biotechnology start-ups encounter. While it's risky to make broad assumptions around the scale-up process, the best approach to optimizing the effort is typically being more aggressive on scale-up factor, with campaign time for pilot process following. The highest risk for commercialization involves pilot and demonstration processes that are not fully integrated.

CONTRACT MANUFACTURING

Scaling-Up Through Contract Manufacturing

As companies consider options to commercialize their advanced biotechnologies, they are faced with the decision to either build their own demonstration scale facility or utilize existing contract manufacturing operations (CMOs). There are benefits and pitfalls to both approaches, with the correct decision founded in the goals of the organization and specifics of the process. Having successfully scaled advanced biotechnologies through both routes, the following is a primer on key decision factors, and a deeper dive into the scale-up through CMOs.

Understanding what success looks like – before discussing the details of scaling-up and the best option for a process, let's consider what success looks like. What are the goals of the scale-up activities? Is it to prove the process technology, demonstrate production cost targets, produce materials to seed the market, or all of the above? The options cannot be fairly evaluated without first understanding the key criteria and prioritizing must haves. Experience has shown that nearly every process can usually be scaled-up, the key question is whether it can be done economically to fit the intended market.

Scale-up is more than data – the single largest area of surprise for most ventures scaling-up a process is the conflict between generating the necessary data to demonstrate the viability of the process, and making representative material to seed the market. These are conflicting goals that may result in a significantly longer period of time for scale-up than anticipated. Let me give you an example:

> An advanced biotechnology startup has been developing an innovative biotechnology and operating in a pilot fermenter with associated downstream recovery, and has demonstrated 70% of its fermentation titer goal. The start-up recently projected to its board the titer goal (critical to closing the next round of financing) would be demonstrated within 6 months. The CTO reports that a new variation of proprietary organism is forecasted to meet the titer target when optimized through the pilot plant. Simultaneously, the head of business development reports that the fortune 100 strategic consumer products client loves the material and needs "hundreds of kilograms" to produce product for a consumer test, which if successful, will result in an offtake agreement. Producing the material for consumer testing will require 4 months of pilot operation.

These conflicting demands represent reality for most biotechnologies during scale-up: The need to produce consistently representative material to seed the market, and the need to discover and change the process to meet key technical goals. Can they be done at the same time? Very unlikely. When making innovative changes, some will succeed and some will not, it will likely be months of pilot operation until a new strain is making representative (on specification) material. The requirement for interim scale manufacturing can often take much longer than initially projected, to meet both technical and market requirements.

Look at financials from a cash basis – when performing the economic analysis to compare building a demonstration scale facility versus using a CMO, the analysis should also be done on a cash basis. Technically, building a demo facility could be capitalized and depreciated, the reality is that once the initial scale-up work is complete, it is a sunk cost that may or may not be utilized further. If the cost is capitalized in the economic analysis, building a facility will appear to be the low cost option, however the reality of scale-up involves a fixed amount of cash and whether it goes towards a capital asset or contract manufacturing does not matter. It is not a P&L statement that is a concern during scale-up, but rather "money in the bank."

Advanced Biotechnology versus Pharma – in "strategic approach to scale-up" I introduced the dollar test, which provides a basis to differentiate products from various markets (pharma, chemicals and food) and outlines the economically viable options for each. The upshot being that for products such as pharmaceuticals, practically any biotechnology option is available, but foods and commodity chemicals have economics that limit the practicality of many technologies. This applies directly to the world of CMO options. Most contract manufacturing options within advanced biotechnology are built and operated to pharmaceutical standards and have a cost structure often 5-10 times higher than CMO's built and operated to non-pharma standards. This makes pharma style CMO's an option for initial screening (say 500 – 5,000 liter fermentation scale), but they become economically prohibitive when moving into demonstration or commercial scale (25,000 liter or greater).

Pros and Cons of Contract Manufacturing – The decision on whether to build a demonstration facility or use a CMO is a combination of cost and ability of each to meet the goals discussed previously.

Advantages
- Likely faster time to market if CMO does not require major equipment modification
- Less up-front cash, limited capital investment and cost incurred as CMO operates
- Benefit from experience of others, not required to be experts on all unit operations

Disadvantages
- Less control, inability to control schedule and specific way equipment is operated
- Higher marginal production cost and higher overall cost for lengthy operations
- Hidden costs of staff time and travel required to provide on-site supervision

The CMO world is getting smaller – most contract manufacturing operations for advanced biotechnology were not built with the intent to be a CMO, they ended up in this mode of operation due to loss of their primary product. This is not a bad thing, but since the facility was not likely built to be a CMO, it likely does not have all the flexibility required. The second reality is that since there have not been as many facilities built in recent years, the available world of CMO's for advanced biotechnology companies continues to get smaller. Contract manufactures seek long term (multi-year) manufacturing arrangements, while biotechnology startups often seek a very discrete and less predictable operational period.

Using a CMO for process scale-up can be the lowest cost option when the time period is shorter (typically 1-2 years max) and the facility can host the process without significant new equipment or investment. Scale-up periods of longer than 1-2 years or that require significant capital equipment installed at the CMO will typically be more favorable towards building a dedicated demonstration facility.

Contract Fermentation Landscape

Funding for early stage advanced biotechnology is ramping up and companies commercializing fermentation-based technologies are faced with the decision to either build their own demonstration-scale facility or utilize existing contract manufacturing organizations (CMOs). There are clear benefits of using CMOs, however conducting a successful CMO search is much like preparing for a difficult hike, the most important factor going in is having the right mental perspective to understand the challenge that lies ahead.

It is important to understand what a CMO is and is not. In the world of biotechnology, a CMO is a manufacturing facility with aseptic fermentation and downstream recovery to make a range of products from microbes. Implied in the name is that these are primarily manufacturing facilities that have long term agreements to make products (primarily pharmaceuticals) from processes with an operational history and have successfully been scaled-up. By contrast, most emerging biotechnology companies looking to do scale-up work, are seeking shorter term arrangements (as low as just a few fermentation runs) to prove their technology and produce samples for potential clients. While CMOs can successfully support scale-up work, it is usually a small subset of the CMO world and finding the right partner is a complicated and time-consuming journey. Some lessons-learned based perspective is as follows:

Don't be a "5" looking for a "10" – sounds like advice you might get from a buddy in high school, but also applicable to most CMO searches. Many early stage biotechnology startup ventures often believe its scale-up project is much more attractive to the CMO world than the CMO world views the work. They get funding and expect a line of CMOs to form – in reality the line doesn't form, calls are answered, and the searches are going global. Pharmaceutical production is strong and the world of CMOs has continued to diminish. Making the effort to find a host CMO is longer and more costly than often expected. Scaling-up through CMOs is still a valid approach, but understanding the market conditions is necessary.

Clearly define your goals – scale-up has many facets, including getting data on fermentation and unit operations at larger scale, generating sample product for potential clients and in the case of novel foods, producing representative materials to be used in regulatory (GRAS) testing. All of these are goals are important, however they often conflict. The client and regulatory product samples are an example. In the case of client samples, the largest amount of product practical is usually the goal, while in the case of regulatory samples, there is often a need for the material to come from a series of batches, not just one large batch. This dictates a need for more runs in smaller fermenters than the product runs. These factors will have an impact on selection of a CMO site and needs to be understood up front.

Start at the end and work backwards – building on the goals discussion, it is important to start with the goals as an end point and work back to understand what is required. What type of fermenter is required? How much downstream recovery? Is a spray dryer required? Does the product need to be produced to food standards? These are just an example of common and critical early stage questions that will reduce a theoretical starting list of 20 or 30 sites down to just a few very quickly. As time and money are usually in short supply for biotechnology ventures, this triage at the front end of the process will save significant time and effort.

Commercial Structure – another key issue to understand up front. There are typically different structures used for scale-up work at contract manufacturing facilities, generally broken into fee for service and hybrid with some form of equity. While traditional CMOs are typically fee for service, they are not always open to shorter duration scale-up activities. The facilities that are often best suited for scale-up work, are often interested in "a piece of the action" in the form of equity. Neither is right or wrong, but experience has proven any deal involving equity, will be much more complicated and have a much longer timeline.

Get a passport – often a surprise to many doing a first time CMO search is the limited number of options within the United States, especially when looking for larger scale operations (10,000 liters or greater) open to supporting scale-up work. Sites in Europe and Mexico are the most often used, with options in eastern Europe becoming more common.

Downstream recovery, where the wheels can come off – as advanced biotechnology is typically driven by aseptic fermentation, this can receive a disproportionate amount of attention during the initial search efforts, but the requirements to make a final product dictate that the CMO have (or be able to assemble) the required downstream recovery. This is commonly a major obstacle in the selection process. While the CMO may have some key equipment, it is uncommon for a

CMO to have all required equipment. The most common approach is to bring in skid-mounted rental unit operations where possible. Most CMOs are set to accept process skids and this is generally the cleanest approach. In the case where this does not work, it is possible to ferment at one site and do downstream recovery at another, but the risks involved in packaging and shipping fermentation broth make this a less desirable option.

Capital Light Scale-Up Approach

Capital light scale-up utilizes existing fermentation facilities, in lieu of purpose-built demonstration facilities, to bring advanced technologies to market faster and with less capital burn. It is a great plan if it can be executed, but risk comes in the form of limited options, making it far from a predictable pathway. With record funding, an impressive backlog of technologies are in early-stage commercialization, but the diminishing landscape of contract manufacturing organizations (CMOs) is setting up a potential perfect storm of demand significantly exceeding supply. The capital light approach can be utopia, but go in understanding the challenges. The following outline some key considerations to ensure capital light scale-up can become a reality.

Pick realism over optimism – Optimism is abundant within advanced biotechnology and it has the beneficial effect of keeping many of us in the fight, through a long and at times, challenging journey. The downside of optimism is when evaluating capital light scale-up with CMOs, where the total cost of the program is driven primarily by how long the operations will take to reach required goals. If funding is based on optimistic assumptions that do not materialize during scale-up, it places the venture in a very difficult and undesirable position of running out of resources to reach commercial viability.

Economics 101 – The economic downturn of 2008 impacted all industries, including contract manufacturing. Little new capacity has come online since then, at the same time many existing CMOs ceased operations or converted to captive use. The demand by advanced biotechnology diminished during the same period, while most ventures retrenched into development mode. The recent surge in funding and organism-focused companies has produced a significant pipeline of industrial biotechnology processes, all beginning the march to commercial operation. This suggests a future of demand continuing to exceed supply, but at this point, it is only a forecast.

Chicken or the Egg – the classic quandary of which could have come first. Even if confident of the coming surge, it is hard to raise capital to expand CMO capacity as the need is only forecasted. Personal experience has shown, non-binding partnership letters will only go so far and usually will not secure large-scale project financing. This sets the advanced biotechnology industry up for big game of "chicken" on who will pull the trigger to expand capacity. As we all learned as kids, the hardest part of playing chicken, is knowing if you need to flinch.

See the whole chessboard – Limiting your scale-up options to existing CMOs is the most defined pathway, but also the most crowded. The larger the scale and longer the need of operations, the wider to cast your net during the search. Idled facilities, brownfield sites and manufacturing facilities with underutilized assets can be a viable, although more complicated option. Looking back at the industrial biotechnologies that have successfully scaled-up over the last 15 years, many used fermentation facilities that were not considered traditional CMOs at the time and show a pathway that needs to be seriously considered.

Apples and Oranges – the process of evaluating CMO scale-up options is a complex exercise comparing facilities with different capacities, costs and capital investments. The primary driving factor is fermenter size and efficiency, which dictates the capacity. Most CMO's have robust fermentation with limited changes required. Downstream recovery is another story, where some facilities will have more of the required equipment than others, compelling rental equipment or equipment purchase. This will generate a cost structure with CMO manufacturing fees, third party rental costs and purchase of equipment, for options with different levels of production. Depending on whether the results are viewed as average $/kg of product or total cash burn for the effort, may generate different decisions from the analysis.

Time is not on your side – for large scale industrial fermentation (50,000 liter fermenters or larger), it can commonly take 6 months minimum to select a site, negotiate a contractual agreement, transfer the technology and do preliminary small-scale runs to be in a position to run batches in a large-scale fermenter. If the CMO requires additional equipment or the business arrangement is more complex than standard fee for service (equity, incentives, etc.), the timing can extend to 12 months or longer. Ventures that have a promising organism and potential product do not want to be the one standing when the music stops. Companies relying on CMOs for product commercialization should start early to ensure they meet public commitments.

COMMERCIALIZATION

Biotechnology Commercialization Roadmap

As the next wave of first-of-a-kind biorefineries ramps-up, the subject of project engineering once again comes to the forefront. Questions about the overall process, why it takes so long, and costs so much abound, all capped off by the realization that an EPC solution counted on is not always the reality. An understanding of the entire process is key for those heading down the commercialization path. The graphic at the end of this section is a _concept to commercial operation_ roadmap, with detailed descriptions that follow.

Project Development
Front-end loading (FEL) is the engineering development process to take a concept from the idea stage to the point that funding sources (banks or governmental agencies) and/or company boards of directors can make decisions to move forward on the project. It is structured in a stage-gated process, with each step targeted at developing adequate information to make an informed go/no-go decision on the path forward. The process is typically broken into 3 phases (FEL1, 2 and 3) that collectively represent what is referred to as Front End Engineering Design (FEED). This process usually costs 2-3% of the total cost of the project and needs to be funded by the company before project financing can be secured.

Execution Phase
The execution phase is the period from the start of detailed engineering to the end of construction (mechanical completion). Detailed design converts the concepts of the FEED package into documents that can be used by contractors to build the facility. Major focus is on developing drawings and specifications for civil, structural, mechanical, architectural, electrical and control systems. Construction begins with the site being prepared, utilities are brought in and foundations prepared. Equipment is placed on foundations and connected with pipes, power and controls. The construction phase is complete when mechanical completion is reached. This is the point where the facility has been constructed, but not tested or operated.

Commissioning and Start-up
Once the plant reaches mechanical completion, the process of commissioning and startup begins. This is a structured process that verifies the operability of each component before operating them individually, then in groups, then as a full process facility. Start-up is the point when raw materials are introduced to the process with the intent to make product.

The technology commercialization process can appear very structured, but is based on decades of best-practice learnings from the process industries. Some key perspectives, applicable to industrial biotechnology commercialization, are as follows:

FEL is different than facility construction documents – The development stage of the project is more strategic than operational, with the goal being to set the framework that the project can be built within. It's focused more on the operational specifications of the process (temperature, pressure, fermentation time, etc) than the equipment details (how thick is the tank wall, what type of concrete foundation). I often describe this as the "interior design" phase of the project, with construction being when the "sheet-rockers" start their work.

The world of engineering firms that are good at FEL is getting smaller – As an alumni of the engineering services Industry, I can attest to the fact that few firms do both FEL development work and design/build very well, as they are dramatically different. The development phase is where multiple options are considered, tweaked and morphed into a final project concept. By contrast, detailed design is a very structured process, guided by governmental regulations and requirements. The slowdown in all process industries over the last 5-10 years has impacted the number of firms that have extensive FEL expertise and this is often the more challenging support to find.

EPC is not what you think – Quite simply, EPC stands for Engineer, Procure and Construct, the primary phases of project execution. Unfortunately, over the years it has become synonymous with an integrated "turnkey" project execution, hiring a firm that can build a fully operational facility and guarantee its performance. This has been common in industries with fully proven technology, such as corn ethanol, but is rare to non-existent in emerging technologies. You can find qualified firms capable of both engineering and building a facility, but unless you have other operational facilities that have passed a performance test, it is very unlikely that any performance guarantee will be provided.

Understanding engineering industry terminology and typical path forward is a key to ensuring the process meets time, quality and cost expectations.

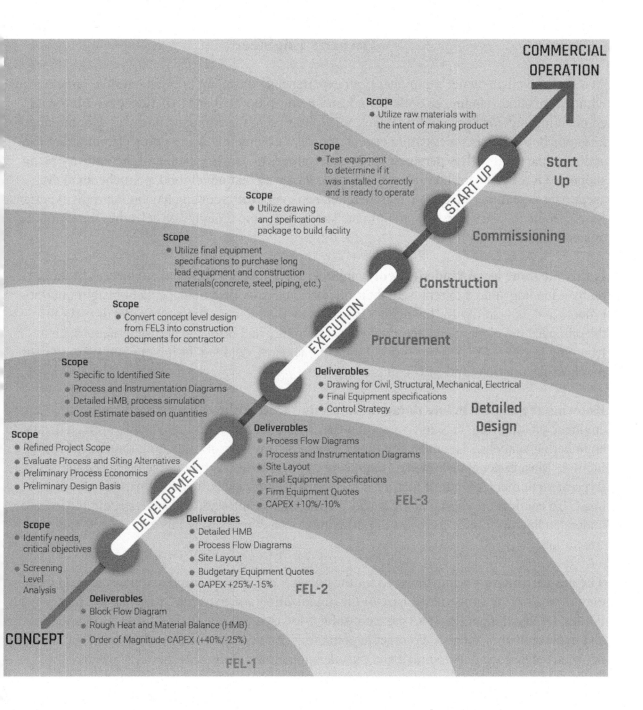

COMMERCIAL OPERATION

Scope
- Utilize raw materials with the intent of making product

Scope
- Test equipment to determine if it was installed correctly and is ready to operate

START-UP

Start Up

Scope
- Utilize drawing and speifications package to build facility

Commissioning

Scope
- Utilize final equipment specifications to purchase long lead equipment and construction materials(concrete, steel, piping, etc.)

Construction

Scope
- Convert concept level design from FEL3 into construction documents for contractor

EXECUTION

Procurement

Scope
- Specific to Identified Site
- Process and Instrumentation Diagrams
- Detailed HMB, process simulation
- Cost Estimate based on quantities

Deliverables
- Drawing for Civil, Structural, Mechanical, Electrical
- Final Equipment specifications
- Control Strategy

Detailed Design

Scope
- Refined Project Scope
- Evaluate Process and Siting Alternatives
- Preliminary Process Economics
- Preliminary Design Basis

DEVELOPMENT

Deliverables
- Process Flow Diagrams
- Process and Instrumentation Diagrams
- Site Layout
- Final Equipment Specifications
- Firm Equipment Quotes
- CAPEX +10%/-10%

FEL-3

Scope
- Identify needs, critical objectives
- Screening Level Analysis

Deliverables
- Detailed HMB
- Process Flow Diagrams
- Site Layout
- Budgetary Equipment Quotes
- CAPEX +25%/-15%

FEL-2

Deliverables
- Block Flow Diagram
- Rough Heat and Material Balance (HMB)
- Order of Magnitude CAPEX (+40%/-25%)

CONCEPT

FEL-1

Owner's Engineer

At a basic level, an owner's engineer is an experienced consultant brought onto the project team with a deep understanding of both the industry being supported and experience working with engineering and construction firms, thus able to represent the company interest. Commonly, it is someone who has successfully executed similar commercialization projects multiple times and joins the team with experience of the entire process. Understanding of the industry being supported is a key factor not to be overlooked. The way that novel food, petrochemical or aseptic fermentation process are commercialized is very different and having the right experience for a specific industry is critical to ultimate success. Some thoughts on selecting the right owner's engineer is as follows:

Is this the same as an "independent" engineer – No. An independent engineer is brought on by a funding source to represent their interests and provide an independent insight into the process and its viability for financing. By contrast, the owner's engineer works directly for the company developing and deploying the technology and brings expertise in items such as front-end loading (FEL), process development packages, detailed design drawings, equipment purchase negotiations and EPC contracting.

Knowing the answer before doing the analysis – One attribute that the right owner's engineer should bring, is to have a good idea of what the answer will be before doing a detailed engineering analysis. The most common example is capital cost estimating for a proposed facility. Most experienced owner's engineers can hear a description of the proposed facility (type of equipment, output of facility, etc.), and will start calculating a ballpark cost in their head. To confirm the capital number will require significant work and analysis, but having an understanding up front if there is a big difference with expectations can save money, time and frustration.

The ala-cart menu – while the need for the owner's engineer is common, the exact role they play depends on the knowledge base of the organization and gaps that exist. This can vary from limited technical input, to a part-time consultant focused on specific areas such as detailed design and management of the EPC contract negotiations. The common theme is acting as the interface between technology development and outside engineering firms or technology providers.

Part of the team – while there is usually an arm's length relationship with traditional engineering firms and technology providers, an owner's engineer usually is integrated into the project team more like an employee than a third party. This includes participation in internal planning and strategy meetings, often resulting in them being the key communicator with outside parties on issues that relate to engineering and construction.

Ability to "phone a friend" – one significant benefit of using an experienced owner's engineer is the network that they bring with them. There is great value in being able to pick up the phone and when questions are slightly outside their area of technical expertise and get input of strategic value from contacts in their network.

Don't get too far out over your skis – if you have read the previous sections, you will recall I have touched on the misunderstanding many in advanced biotechnology have as to what EPC is and is not. It becomes important, as so many early stage ventures worry about the strategy and partners needed to build their second or third plant, they do not pick the right firm to design their demonstration facility. I strongly recommend picking the right engineering firm for the task at hand and have found it to have a better record of success.

Understanding engineering industry terminology and typical path forward is a key to ensuring the process meets time, quality and cost expectations.

Bioprocess Capital Costs

Discussion of biotechnology capital costs is often a controversial topic, on par with discussing politics at the Thanksgiving dinner table; why the facilities are so expensive and a disbelief of the forecasted costs. As a seasoned veteran in the battle, trust me when I tell you that the cost of a fully-constructed industrial biotechnology facility will be about 3 times the purchase cost of the major equipment. Similar to election day, just because the estimate is not the desired outcome for some, does not mean it's not reality. Let's explore:

Back of the envelope – In the early phases of project development there is a need to understand the order of magnitude capital cost of a proposed project. Quite simply, knowing whether a proposed facility will cost tens of millions or hundreds of millions will make a big difference in process development and fundraising efforts. This is accomplished by factor-based estimates, determining the major equipment required and multiplying by industry factors (known as Lang factors) to determine an order of magnitude costs. They are determined from previously executed projects and compare total project cost to the cost to purchase major equipment. If a previous project had a total cost 3 times the cost of equipment purchase, it is a reasonable assumption a similar project will be 3 times the equipment cost.

It's a secret – The factored estimate approach seems easy, if you have accurate cost factors. The problem is, for emerging industries like industrial biotechnology, there is not a lot of publicly available documentation supporting them. You can often find them in public documents for more mature industries like pharmaceuticals (5 to 6 times) or petroleum refining (4 to 5 times), but not for emerging industries. I can tell you from personal experience that the Lang factor for industrial biotechnology range from 2.5 to 3.5 times the equipment cost and this is generally accepted by engineering firms and capital project groups, but unfortunately all the data to support this is based on proprietary information. NREL has published a series reports on biorefinery economics that calculated detailed capital cost estimates and backs into Lang factors. While they are for proposed facilities and not actuals from completed projects, they are one of the best publicly available sources to support the range.

It just can't cost that much – a standard response I have received many times over the years by senior management who sincerely do not believe it can cost more to install equipment than to purchase it. As noted in the paragraph above, installation factors from other industries are ABOVE the range I quote for industrial biotechnology. While the hopes of a plant only costing 50% more than the equipment (1.5 installation factor) is an admirable dream, it is unfortunately, a dream. The installed cost ranges for individual project components will vary, but for most industrial biotechnology projects, the chart below illustrates typical cost percentages for primary project costs. While equipment purchase is a major component, it only represents about one-third of the total project cost.

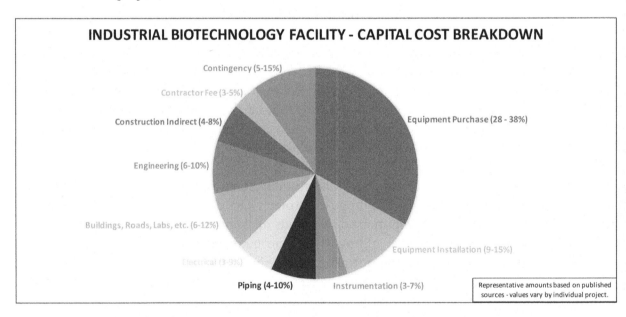

Everything I need to know I learned in ChemE 201 – The first and most basic chemical engineering class at most universities is material and energy balances, developing detailed analysis of what compounds go where in a chemical process; and how much energy is added or removed along the way. It is the key starting point for any capital cost estimate and provides required project definition. Think of it this way, you cannot determine how much a car costs until you define what kind of car you plan to buy. Taking shortcuts in this area is a mistake, but a well-executed effort pays dividends down the road.

Sizing the equipment is key – When developing a factored capital cost estimate, the most important item is to determine the size (and thus cost) of the equipment. The key to doing this accurately is understanding flow rates, required separation efficiencies and working with vendors to select the right equipment. Data from the specific process is critical and should not be "assumed". Guesstimating the equipment size will generate a factored cost estimate that provides limited value and inadequate accuracy to be relied upon.

Understanding the basic principles outlined above will provide a perspective that allows early-stage capital cost estimates to be used for key strategic value and not as an aspirational (and often unachievable) target.

Biotechnology Manufacturing Cost Drivers

As advanced biotechnology begins to ramp up process development efforts and deploy the next wave of first-of-a-kind biorefineries, the subject of accurately forecasting commercial facility cost drivers comes to the forefront. Understanding what will make a technology economically viable is key to focusing development efforts to areas that will drive ultimate success. Having the opportunity to support many technologies path to commercial operation has provided the following insights from both the benchtop looking forward and commercial operation looking backward:

Breathing your own exhaust is harmful to company health - Optimism is abundant within advanced biotechnology and it has the beneficial effect of keeping many of us in the fight, through a long and at times, challenging journey. The downside of optimism is when evaluating the reality of economic challenges. Raising capital usually involves putting forward a favorable economic model and sales pitch supporting why it is achievable. I often advise investors that a proposed plan is "possible, but not probable", meaning it can be achieved, but is far from a clear path to success. It is hard for companies in this mode to retreat to fairly assessing the challenges that lie ahead for the technology. The challenge is best summed an up by an industry executive upon finding out that technical challenges were ignored, "Just because risks are assumed away in a financial model, does not mean they actually go away."

Focus on what matters – techno-economic modeling is a key offering of my consulting practice and clients are often surprised at how streamlined my models are. I would argue that a good first pass can be done with a calculator and whiteboard, as the key criteria are easily identifiable. My approach is simple for a reason, once you get past the top 5 costs factors (product yield on carbon source, labor, depreciation, media costs and utilities for example), not much else matters in the early phases of development. These typically represent over 90% of manufacturing costs and significant efforts at modeling minor variations is usually a waste of valuable resources. Focus on the big items and factors that make a material change in then.

Challenge historical perspectives, but don't ignore them – Working early stage biotechnology over the years provides perspective. First, it's common is to challenge the status quo, which I fully support. Change often should happen, but it requires a plan backed by a solid technical approach. If the reasons for taking a non-traditional path cannot be clearly articulated and justified technically, pick another area to change.

Benefits of building large facilities – One of the most common lessons learned from the last round of biotechnology commercialization was that companies built too large and too fast. While I agree, there were reasons for going big that need to be understood. Building smaller plants requires less capital, but brings with it inherent production cost disadvantages shown below. Labor is the most common example. While a commercial facility often has production levels 25-50 times larger than its demonstration scale equivalent, it would normally only require 5-10 times as many staff. The end result, much lower unit costs as the capacity of production increases.

Fermentation Product Manufacturing Costs
Typical Protein Product

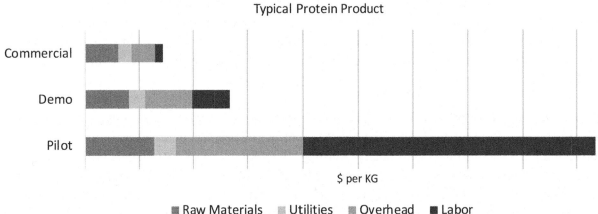

$ per KG

■ Raw Materials ■ Utilities ■ Overhead ■ Labor

Downside of being sensitive – Detailed sensitivity analysis of early stage financial models are often not a good use of resources. I have seen significant workups with waterfall charts and tornado diagrams for models that have an overall accuracy of +50%/-25%, where the individual components do not vary within the accuracy of the analysis. It at times has reminded me of a cat chasing a laser pointer, tons of activity, but no tangible result. Meanwhile, limited resources are spent on the technical issues like yield improvement, that drive the highest impact on cost. This is a point where the 90/10 rule applies, the vast majority of the early resources should be placed on verifying and supporting the key technical assumptions that go into the economic model and limited effort towards the deep sensitivity analysis.

Make sure the levers you plan to turn actually work – reviewing many models over the years, I am familiar with the concept of levers, technical inputs that can be changed to improve performance. Key items like alternate feedstocks, byproduct valuation and product specifications that will allow for improvements to the economic model if performance is less than projected. Looking back from the perspective of having built and operated commercial scale facilities, I can advise that many of the economic levers assumed in the early stages do not turn out to be an option at commercial scale. Early work to identify factors that will have a material impact on ultimate facility operations is effort well spent.

Following the basic principles outlined above will produce techno-economic models that more accurately represent commercial reality and improve probability of success.

Determining Facility Capital Costs

How much a facility will cost to build is often a topic of hot debate and there are many industry standard methods to calculate it. They have increasing levels of effort depending on the level of accuracy required. Generally, estimates are broken into three categories outlined below. The details of the FEL phases will be outlined in a later section:

Estimate type	Level of Engineering	Accuracy
Order of Magnitude	FEL-1	+40% / -25%
Feasibility Level	FEL-2	+25% / -15%
Authorization Grade	FEL-3	+10% / -10%

Order of magnitude estimates are developed two ways during the FEL-1 phase. The first method involves factoring based on a facility of similar size. In the event you have capital cost information for a facility, you can calculate a facility of similar size by multiplying the base cost times the ratio of the new capacity over the old capacity to the 0.6 power. An example below.

Your company builds a facility to 1,000,000 pounds per year of your product for $12,000,000 and are looking to build a facility ten times larger, producing 10,000,000 pounds per year. The cost of the new facility is:

$12,000,000 x (10,000,000 pounds/1,000,000 pounds)^0.6 = $47,772,860

Unfortunately, for new processes, reference facility information is not readily available, and estimates are done by a factored approach. This involves calculating the cost of the key process equipment for the facility and multiplying by an installed cost factor as shown below.

Direct installed cost = (purchased equipment cost) x (multiplier)

These multiplication factors are developed from many years of studying installation of typical process equipment and when done on a plant level basis are referred to as Lang factors. For most industrial biotechnology processes, the total installed cost of a facility will typically be 3 to 3.5 times the cost of the equipment. If you estimate $10,000,000 of equipment, an order of magnitude cost of the facility will be $30,000,000 to $35,000,000.

The purchase cost for a given component can be developed from direct vendor budgetary estimate (if available) or approximated from estimates for similar equipment from other projects. To adjust the price based upon differences in equipment capacity, the estimated purchase cost is developed by what is known in the process industry as the "rule of 0.6". This formula adjusts the estimated purchase price based upon the ratio of the equipment new capacity to base equipment capacity to the 0.6 power:

New Cost = base cost * (new capacity/base capacity)^0.6

Feasibility level cost estimates are developed during the FEL-2 phase and are generated by using equipment specific installation factors. Where the plant level approach uses one average factor for the entire facility, each type of equipment has a different installation factor that typically range of 1.5 to 4 and are based on the difficulty to install. Items like skid mounted equipment are on the lower end of the range and specialized equipment like fermentation are on the high end of the range. There are various public sources available including Sterns Catalytic.

To develop a cost estimate once all the equipment is summarized and multiplied by the appropriate installation factor, indirect costs including engineering, construction management, contractor fee and contingency are applied to reach a total installed cost estimate. Cost estimates based on this rationale generate a feasibility level facility cost estimate with an accuracy of +25% / - 10%.

Appropriations grade estimates are developed during the FEL-3 phase and are customary for receiving authorization from a company board of directors or debt funding source. This type of estimate has an accuracy of +10% / -10%. This involves a much more rigorous level of engineering design than the previous estimation approaches and is based on unit quantities, a determination of feet of pipe, yards of concrete, number of control loops and other key cost items. These estimated quantities are multiplied by unit costs to determine the construction cost. The basis of equipment costs for an appropriations grade estimate is also more rigorous. While early stage factored estimates are done from budgetary estimates (rough estimate based on similar equipment on similar projects), firm executable cost estimates are the basis required and entails developing a detailed specification for each piece of equipment and obtaining binding quotes from vendors.

Project Execution

Industrial scale biotechnology projects are complex ventures that require significant time and capital resources to execute successfully. This summary provides an overview of the project execution process, broken into the following phases:

- Front End Loading (FEL)
- Project Delivery
- Commissioning and Operations

The document also provides typical project and cost timelines. It should be noted that the time and cost examples are based on key assumptions that will require confirmation for the specific project, however can be useful in the early phases of the project for planning purposes.

Overview and Project Development

When I reflect on the characteristics of successful advanced biotechnology companies that have achieved commercial scale, and the longer list of companies that haven't, the difference between success and failure is often not the core technology alone, rather the lack of knowledge and execution of navigating through the commercialization process. Specifically, the navigation from process concept and engineering stages, through construction and startup, to the pinnacle goal of a reliable and profitable manufacturing operation. Reaching that reliable manufacturing operation is the goal our industry cannot lose sight of, to succeed as an advanced biotechnology community, we need more technologies to reach commercial scale and endure long-term manufacturing operations.

The overall commercialization process, from concept to commercial operation for a first-of-a-kind commercial scale biotechnology, as illustrated below:

PROJECT EXECUTION TIMELINE

Timeline

This following summary is intended for anyone who is working to take a new technology from the bench/pilot stage to commercial operation. This includes engineers, scientists, management and investors, anyone who will benefit from a detailed understanding of the process before they begin. Think of this way, if you are heading out on an 8-mile hike, wouldn't it be good to know if there is a bridge washed out at the 6-mile mark? The same can be true of commercialization. You do not want to spend significant resources to prepare a bank finance or federal loan guarantee application, to find out your technical package is insufficient. The process is outlined as follows:

Front End Loading (FEL) Stages

The purpose of the FEL process is to develop a process concept to the point of definition that a board or financing entity can give approval of funding to move into project execution. Often alternatively referred to as Front End Engineering Design (FEED), is defined by three front-end loading steps (FEL 1, 2 and 3) defined below:

Front-End Loading (FEL) 1: Business Planning
The primary objective of the business-planning phase is to define the business opportunity. After a proposed project has been identified, it must undergo a certain amount of definition. This includes a technical assessment, development of a milestone schedule, and an estimated cost range (typically a +40/-25 percent estimate). As the proposed project becomes better defined, a clearly formed business plan can be framed. Concurrently, economic models and business evaluations can assess the proposed project's strategic importance and its business, production, and financial potential.

Front-End Loading (FEL) 2: Scope Development
This stage is referred to as the scope development phase. The beginning of this phase is the formation of a project team that will start to develop both the business objectives and project scopes for a given business opportunity. A Project Manager will typically be assigned to manage the FEL 2 development and into later stages.

During the FEL 2 stage, teams develop multiple alternatives that meet the business opportunities. However, it is important to understand that at the end of FEL 2, the project gatekeepers must identify and choose only one as a facility design basis of the alternatives to develop during FEL 3. In addition to selecting one alternative, it is important that the following are outlined for the final alternative chosen to allow the project to utilize the next stage effectively. Work products generated in FEL 2 are the project objectives and conceptual engineering package outlined as follows:

- Site assessment and selection
- Site ambient conditions, topography, soil conditions, design codes
- Site Plan
- Process flow diagrams containing basic control concepts with utility (average and design) loads
- Process and long-lead component list
- Draft P&IDs for key systems
- Project schedule
- Budgetary estimates for major equipment obtained from vendors or based on historical data
- Total installed capital cost estimate (+25/-15 percent) "CAPEX"

Front-End Loading (FEL) 3: Project Planning
This stage is referred to as the project planning stage. The beginning of this phase is the point at which one alternative evaluated during FEL 2 has been selected for further definition, with the goal of taking it to an authorization board for funding. During this phase, most project teams grow in size due to the increased amount of engineering work to be completed prior to authorization.

The goal of FEL 3 is to develop a set of engineering documents (design basis package) that incorporate site-specific conditions and a plan for executing the project, such that reliable cost and schedule estimates can be established. The FEL 3 stage CAPEX typically reflects an accuracy of between ±10 percent accuracy. The work product of this phase will allow a detailed package to be presented at the authorization gate. The specific deliverables for the FEL 3 stage are updated FEL 2 deliverables, plus the following:

- Complete P&IDs
- Process plant area equipment layouts (plan and elevation)
- Detailed equipment technical specifications for major process equipment, suitable for procurement
- Purchase ready bids for major equipment
- Project execution and procurement plan
- Detailed scope of work (including "bulk" quantities and material takeoffs)
- Valve, pipeline, instrument and cable list with quantity estimates
- Architectural layouts and plans, including labs, offices, control room and other buildings.
- Critical-path method, resource-loaded schedule
- Preliminary project commitment, cash-flow and budget
- Authorization-grade estimate (±10 percent accuracy)

The end of FEL 3 occurs when the project is authorized and the project team receives funding to move into detailed engineering. The table below summarizes the overall FEED process into its three components, describing the scope of each effort and providing a typical duration, cost and accuracy of capital estimate. These estimates are for full commercial scale projects. Demonstration scale projects will follow a similar process, however the timelines would normally be reduced.

Phase	Description	Key Documents	Duration (months)	Capital Accuracy
FEL-1	Business planning stage to define the concept	• Block flow diagram • Rough material balance • Order of magnitude cost	1-2	+40%/-25%
FEL-2	Scope development, early phases of engineering and budgetary equipment estimates.	• Process flow diagram • Material and energy balance • Process layout/site plan • Equipment specifications • Budgetary equipment quotes	2-4	+25%/-15%
FEL-3	Process planning and development of basic engineering package	• Process and Instrumentation diagrams (P&ID's) • Detailed site plan • Firm equipment quotes • Detailed utility estimates • Quantities of pipe, concrete, etc	4-10	+10%/-5%

FEL 3 is the level of detail required
For a bank loan or federal project
loan guarantee (DOE/USDA)

Project Delivery

Once funding is secured for the proposed project, it moves into the delivery and execution phase. This is an exciting time for a new technology company, as it is the point when a technology startup is given the green light to turn their process into steel in the ground. While exciting, it is also a make or break moment.

The best analogy I can give to represent the challenges and rewards of the new biotechnology commercialization process is watching the movie Apollo 13. In the movie, after having major mechanical issues, the crew is facing the challenge of re-entering the earth's atmosphere. There is a long discussion of the re-entry window and the sliver of angle that allows for a safe return. Try to re-enter the atmosphere at too shallow of an angle, and you will bounce off; come in too steep and you will burn up in the atmosphere. It is only those who can hit the narrow re-entry window that survive and prosper. This is ironically similar to new biotechnology commercialization. Commercialize too aggressively and you may start up a money losing operation that burns through all of your cash. Study the concept for too long and you will run out of cash before you can bring the venture online. It is the proper balance of lean-forward aggression and smart risk-based decisions that allows certain ventures to reach commercial operation.

The amount of funding secured for project execution is often equal to all of the funding burned by the organization to date, and will ultimately be routed through a large construction firm to execute the project. Understanding how to make the selection of the right engineering and construction firm, along with the process that will follow, is critical to gaining commercial success.

Execution

The execution phase is defined as the period from the start of detailed engineering to the end of construction (mechanical completion). During this phase, the project team is converting the basic engineering package that was developed for the authorization estimate and schedule into detailed engineering drawings that will be used for construction of the project. The execution phase is considered complete when the project is mechanically complete, and the project team is transitioning the project to the startup team members. The manner in which the design and construction is performed can range from cost reimbursable engineering and construction, to an integrated design/build contract with cost and performance guarantees. While this is more attractive for bank funded projects, it typically increases cost and timeline. A summary of the various contract mechanisms and their attributes is as follows:

Mechanism	Description	Advantage	Disadvantage
Cost Reimbursable (aka time and materials)	Engineering and construction done on a rate sheet	Maximum flexibility, owner control	No maximum project cost. All risk by owner
Engineer, Procure, Construction-Management (EPCM)	Engineering time and materials, fixed fee construction	Owner maintains control, sets incentives to control costs	Project cost not capped, requires detailed design
Fixed Price Construction	Fixed price for engineering and construction	Project construction cost is guaranteed, but not performance	Contractor orders equipment, owner has limited control
Fixed price construction with perform. guarantee	Contractor provides turnkey project with cost and performance guarantee	Project cost, schedule and performance are guaranteed.	Highest cost and limited ability of owner to make changes.

Detailed Design

This phase converts the concepts of the FEL3 package into documents that can be used by contractors to build the facility. Major focus is on developing drawings and specifications for civil, structural, mechanical, architectural, electrical and control systems. This work will typically be conducted by an outside engineering firm that may or may not be part of the construction firm.

Construction

Phase when the facility is constructed. Site is cleared and graded. Roads are cut and building are erected. Utilities are brought onto the site. Piling is driven as needed and foundations are poured. Equipment is placed on foundations and connected with pipes, power and controls. The construction phase is complete when mechanical completion is reached. This is the point that the facility has been constructed, but not tested or operated.

The manner in which a contractor is engaged depends on the form of financing (equity or bank debt) and driving force of the company (speed to market). The table below describes the key project components and who retains primary responsibility under each contract mechanism. The more risk that is passed to the contractor, the more control they retain over the project. Worth considering before deciding on a contractual approach.

Mechanism	Project Scope	Project Schedule	Project Cost	Plant Performance
Cost Reimbursable (aka time and materials)	Owner	Owner	Owner	Owner
Engineer, Procure, Construction-Management (EPCM)	Owner	Owner	Owner and Contractor	Owner
Fixed Price Construction	Owner	Contractor	Contractor	Owner
Fixed Price Construction with Perform Guarantee	Contractor	Contractor	Contractor	Contractor

Cost Considerations

Project spending levels will vary over the project timeline, with key components outlined as follows:

- Project Development (FEED) stage – usually about 2% of the total capital cost for commercial scale projects, with FEL 3 being the majority of the cost.
- Detailed Design – will vary given the size of the project, but typically around 6-8% of the total capital cost
- Procurement (equipment purchase) – rough 30% of the total project cost for commercial scale industrial biotechnology projects. Major equipment requires typical payment schedule of 20% with order, 40% progress payment and 40% upon receipt.
- Execution (construction) – has a front-loaded cost for site mobilization, then typically ramps up to peak staffing near the end of construction. This represents the remaining cost of the project, roughly 60% of the total.

Start-up and Operations

The construction phase of the project culminates in a facility that is ready to be started up and put into long-term production. This is referred to as mechanically complete, the point the facility has been constructed, but not tested or confirmed to be ready for production. This is both an incredibly rewarding and challenging time for a new technology company, as it reflects a change from discovery to implementation. It does not mean the company will cease from innovating, but the research function and production operations will be separate entities with very different goals.

This organizational change in focus is often as challenging as its technical hurdles. What makes a good scientist and researcher is a drive for discovery and unwillingness to accept any constraints, other than the laws of thermodynamics. This is often the DNA make up that allow brilliant people to discover concepts missed by others. Most researchers are motivated by the challenge and thrive in a rapidly changing landscape. By contrast, what makes a good production process is predictability and reliability. The knowledge that you have a very high probability of success in making your product on time and at the cost you projected. This comes from structure and standard procedures, where most good operations staff thrive. If these two worlds seem different, they are. Both are valuable and critical for long term success in their own way, but when the worlds collide, it can make for big problems. Making the cultural transition is a must for commercial success.

From a technology standpoint, understanding the detailed process to bring the process online in a safe manner, proving product quality and ramping up to nameplate capacity is critical to the ultimate success of any technology.

Mechanical Completion

Once the plant reaches mechanical completion, the process of commissioning and startup begins. This is a structured process that verifies the operability of each component before operating them individually, then in groups, then as a full process facility.

Commissioning

Commissioning involves testing the equipment to ensure it has been installed correctly before bringing raw materials into the plant. All pipelines are pressure tested. Utilities such as air compressors and boilers are operated. All control valves are stroked to ensure they operate as designed and all motors bumped to ensure they are turning the correct direction. This is a stage of proving system operation without the intent to make product.

Start-up

Start-up is the point when raw materials are introduced to the process with the intent to make product.

NOVEL FOODS

Commercializing Innovative Foods

Understanding the challenge of commercializing innovative foods will separate those that change the future of food, from others that become an asterisk in a market study. In previous sections I expanded on the technical challenges of commercializing advanced bio-technology. While most of the scale-up lessons are directly applicable to the emerging shift to innovative foods, there are other key factors that must be considered to ensure success.

The first place to start is a discussion of what I mean by innovative food. Traditional food companies have made significant improvements over the years in the cost, nutritional value and safety of the foods we eat every day. I do not intend to ignore their progress, but there has been a recent push (backed by robust venture funding) to create a new sector of innovative foods. These are primarily proteins that traditionally come from animal farming, with the goal of producing indistinguishable products from non-animal sources. You can find a plethora of new and exciting companies making novel foods by application of advanced biotechnology such as meat, milk, eggs, cheese, fish, gelatin and other products. There are two common themes among these companies that define the sector, they are based on advanced biotechnology and their measure of success is a product that is indistinguishable from the current animal-based offering. The second item is the key. While there have been alternatives available for many years to most animal-based offerings (veggie burgers, soy milk, etc.) they are not considered indistinguishable. Consumers are aware they are not eating the "real thing", but make a choice for health, religious or other reasons. The primary goal of the innovative food industry is to provide alternatives that do not require this choice, which is both game-changing and a high bar to meet. As I look forward and outline what is necessary for this burgeoning industry to succeed, let's review how most protein products are made within biotechnology today.

The classic route of biotechnology-based protein production is by aseptic fermentation from an organism that can excrete the protein into the fermentation broth. The protein is isolated from the broth by clarification, combined with techniques such as various forms of filtration (UF, MF, TFF) then purified by chromatography. Finally, the protein is ensured to be free of pathogens and undesirable organisms by utilizing a sterile filtration process. This likely sounds familiar to those who have worked in advanced biotechnology, however if this is the starting point for an innovative food process, there are many obstacles with this approach that make it unlikely to reach commercial success in most food applications. The first concern is that while these technologies are robust and very appropriate for applications with high product values (pharmaceuticals), they seldom produce food products that are price competitive. Secondarily and often more importantly, they do not readily fit the regulations and requirements of producing commercial food. Gaining an understanding of the applicable food safety requirements and how they constrain innovative food technology is a critical first step at the beginning of the development process. Otherwise there is a significant risk that the process developed will not reach commercial success.

The following is a summary of the key concepts of making food products and can provide a vehicle for self-assessment. The intent is to outline the difference with traditional biotechnology and why a shift in thought process is required to bring a novel food to market.

"Fizz-Muh" – when I first meet with companies looking to commercialize an innovative food technology, I can often assess the understanding of the task at hand by inquiring about their strategy for complying with the food safety modernization act (FSMA, aka "fizz-muh"). This was a major overhaul of the US food safety regulations in 2011 that governs how food is produced and what is considered safe to eat. For any food producer, FSMA is the tail that wags the dog. An understanding of the requirements of the regulation and how it impacts manufacturing operations is critical, or the company risks going down a long process development path that results in a great product that cannot be sent to market. It is important to note this regulation is different than the generally regarded as safe (GRAS) determination that new food products must make.

The path to safe food – another area that is often eye opening for those moving from traditional biotechnology into food applications is how food safety is managed, and more specifically how organisms are controlled. In pharma applications, the expectation is that fermentations will be clean (no foreign organisms) and the end product sterile, free of any pathogens or foreign organisms. That is not the case with food. If you look at a specification sheet for commercial food products such as flour, you will see that they are free of federally regulated pathogens (e-coli, listeria, salmonella, etc.), but have acceptable levels for other non-pathogenic organisms including mold, yeast and coliform.   Pasteurization is the typical approach to controlling organisms in food, but many novel proteins that wish to maintain functionality will require an alternate method, that still meets the food safety requirements.

Think sanitation, not sterility – one of the distinct differences between traditional biotechnology and food production is the concept of cleanability of equipment versus the ability to sterilize it. In biotechnology, steam or chemical sterilization is the standard. Much of the equipment is rated such that it can be regularly heated to a temperature and pressure to achieve sterility. For cost and practicality reasons, that is not how food equipment is manufactured and thus, not how the food safety regulations are structured. The ability to access and clean equipment, followed by a verification of cleanliness is the backbone of a typical food safety program. This is a case where the technology being used may be robust and a great fit for biotechnology, but it's really a square peg trying to fit into a round hole when making food.

Proactive versus reactive – the question I get the most after reviewing the constraints above is: why does it matter, in the end I can just test my product and as long as it passes all the tests, it meets the regulations, right? Under FSMA, that is not the case. Food safety programs that rely solely on back-end quality testing are not allowed. One of the key concepts of FSMA is that there must be adequate control points throughout the manufacturing process, with quality testing as a second line of defense, not the primary method.

Who's the boss? – in the case of food safety, from a practical perspective, it is the state or local health department. This is the entity that will issue your facility a permit to manufacture food and administer the food safety regulations. FSMA sets the guiding principles that flow down to the local level, but the group permitting and inspecting dairies, meat packers and other traditional food manufacturing facilities (and bringing that history and perspective) is the primary regulator.

The key to success in commercializing innovative foods is to bring all the brilliant people and novel technologies of advanced biotech to food applications, with the understanding they will need to figure out how to modify their square peg to fit into the round hole that is food manufacturing.

Bringing Innovative Foods To Market

Food is a rapidly changing landscape with many emerging technologies; meat without the cow, eggs without the chicken and sushi without the fish to name a few. These products come from advanced biotechnology, deeply rooted in technology used for pharmaceutical production. Key compounds are commonly produced by aseptic fermentation and standards require pure products, with no contamination. Only the target organism can exist in the broth at the end of a fermentation.

Alternatively, traditional food production has been developed from centuries of farming animals and plants, harvesting and processing as needed to allow storage in a manner to be safe to eat. Based on natural processes, food products are not typically pure, containing microorganisms, with items like yogurt having them as a key ingredient. Can these two different approaches emerge into a successful novel food industry?

The primary question is what type of company these ventures are, biotechnology or food? Without question, they are first and foremost a food company. Maintaining this understanding is critical to commercial success. The technology learnings from pharmaceutical and advanced biotechnologies are valuable, but that technology must fit within the food regulatory framework to become commercial. This can be a frustrating journey for many, but can be achieved by having a broad mix of backgrounds involved, including food industry expertise. Some specific lessons learned commercializing novel foods follows:

Don't fight City Hall – the food regulatory approval process can seem foreign and often does not make sense to people with a long history in pharmaceutical production. The GRAS approval process is a prime example. There are many in the novel food industry that assert the process does not make sense and the standards are not appropriate. My advice? Complying with the current regulations is your fastest and most cost effective path to commercial success. If you want your model to be based on "blowing up the status quo" like UBER, make sure you raise billions like UBER has. If your series A round is $10-$25 million like most of the industry, I suggest finding a way to work inside the process.

Food Certification – One of the biggest challenges for novel food companies is to scale their process in a manner consistent with food safety regulations, including production at a facility permitted to manufacture food. Most advanced biotechnology host sites used to scale-up fermentation technologies are not approved for food production, which brings a complexity to the process that needs to be managed. The basic technology can be demonstrated at these sites, but the ability to taste product and provide market samples is significantly hindered.

Fermentation is just the beginning – While fermentation is often the key technology to producing the novel food product, it is usually just one step in an overall process. Much like an engine is key to operating a car, an engine without a body and wheels is of no value either. The ability of fermentation to be matched with the proper recovery and purification process, and scaled-up as a system, is key to a commercially viable operation.

Scale-up options are limited – Early stage technology companies are continuously faced with the choice of building their own demonstration scale-facility or contracting with an existing facility. As most ventures discover, this is a difficult choice between two costly and time consuming options. Building a demonstration-scale facility typically costs tens of millions of dollars, however finding a cost-effective contract manufacturing location can be as challenging. The number of available facilities has been declining in recent years, at the same time early stage biotechnology companies are raising impressive amounts of cash to commercialize their technology.

Scale the technology and the market – Scale-up is an engineering process that demonstrates the technology will work at a larger scale and drives much of the requirements for a demonstration scale facility. What is often missed is the need to seed the market at the same time. Proving a product can be made to meet cost and quality targets is great, but unless the market draw is demonstrated at the same time, the ability to raise the next-level of funding will be limited. This manifests itself if the need for a larger demonstration stage than often forecasted to fulfill both needs.

The key to novel food companies being successful is bridging the gap between biotechnology and traditional food. The emerging novel food technologies are robust and exciting, with the ability to focus development efforts into the existing food regulatory environment key to becoming commercially viable.

MICROBIAL PERSPECTIVE

Scale-Up From The Microbes Perspective

On a biotechnology project years ago, one of the team members responsible for biological advancement referred to themselves as the microbial organisms "shop steward", a reference to union representation. While it was somewhat in jest, it did imply that to get the most from the microbe, its needs had to be taken into account. It's worth looking at scale-up from the perspective of the microbe for some valuable lessons.

It is important for me to start by pointing out I am far from an expert on the biologic side. I have been involved in commercializing many industrial biotechnologies, but my expertise is from the engineering and process development perspective, with only a basic understanding of synthetic biology and genetic modification. That said, I do believe the perspective of seeing the benefits and impacts of the genetic changes, rather than the technology performing them, is valuable and outlined below.

They are not that in to you – it can be a very competitive world for a microbe at a biotechnology start-up. You are selected for your pedigree, get your genetic modifications, show improved performance to what is available today, and what happens? In many cases you get replaced by an even better version within a short period of time. It is the reality of synthetic biology, the ability to make rapid changes and improve organism performance is great. The hard part for companies is to determine when the achievement is of a level that will be difficult to replicate and worth process development around the organism. There is danger in technical staff getting too emotionally vested in a certain microbe that can be replaced over-night.

You never call or write – when a microbe "graduates" from the synthetic biology side and makes its way into the process development part of the business to determine if it is as valuable as initial testing indicated, it is important to have a feedback loop. Often the biology and scale-up groups of the business do not share enough information, to the point where the staff developing the hopeful microbes, do not know how each one turns out. Some companies go so far as to post tracking charts in the lunchroom so that those that feel vested in microbes they have help develop get to track their ultimate success. This feedback loop is a critical part of the development process and can assist in breaking down departmental silos.

Change that matters – from a simplistic perspective, synthetic biology is based on selecting a base microbe with many of the traits desired and making "tweaks" to get the organism to make a different compound or other desired functions. This all sounds benign until you sit in meetings and hear about over-expression and knocking genes out. They sound minor to many, but I have seen unintended consequences occur. As a base organism is developed more and more, it can often change attributes that were not intended. Some are easily discernable (like a strain that became referred to as the "dead body" microbe because of its incredibly foul odor during fermentation) to less visible ones such as impacts on downstream recovery. Change is important, but like with people, it can be much harder to change only the portions you want, without impacting the rest.

Impacts of a slow metabolism – metabolic pathways are often the primary consideration when selecting an organism, which ones can effectively and efficiently do the desired conversion. This is important to complete the first half (fermentation-side) of most processes, but can present issues with recovery, depending on the proposed process. If the product is a whole cell material where limited downstream recovery is required, this may not be an issue. Products that require extraction from the cell are where the issues begin. Factors like ability to deconstruct the cell to access the target compound, while not converting it to a "soup" of cell pieces that requires costly recovery, is a factor that needs to be balanced against metabolic considerations.

Shiny objects – as mentioned above, one of the most difficult decisions a startup biotechnology venture can make is when a new version of the microbe is worth the significant investment that scaling-up will require. This usually involves small-scale bench fermentation to do rough optimization of operating parameters, followed by pilot scale runs in pilot plant or at a contract research facility. As pilot scale runs can often cost tens of thousands of dollars each, the cost to determine if the next hot microbe is a winner or not can be significant. The feedback cycle of learning what forecasts success at an early stage is critical to optimizing scale-up spending.

Synthetic biology is an incredible tool that has already made dramatic improvements to everyday life. The ability to integrate it even further will come from a better understanding of the entire scale-up process, including viewing it from the microbes perspective.

What A Microbe Would Say On Glassdoor

In the previous section, I discussed how the biotechnology commercialization process would be viewed by a microbe. This is a little lighter topic than the majority of the book, but I decided to double-down and ponder further, considering what comments would a microbe post on Glassdoor about their start-up company.

The food here used to be great, but now it sucks – Life for a promising microbe starts out good, every need attended to by highly trained technical staff. The food is top quality sugar, often reagent grade. Then as the company moves down commercialization path, the cost-cutting starts. Buzz words like "cash flow" and "operational efficiency" start flying around and the company quietly transitions to industrial dextrose. That may not be too bad, but before you know it, the microbe is being fed raw cane juice or crude glycerin. Did the microbe really expect to consume thick brown liquid when it joined this company? Something the microbe should have considered and a warning that the awesome coffee service may not be around forever.

The downstream processing group is mean to me – Unfortunately, the cold hard reality within advanced biotech. Once the organism is grown lovingly by the fermentation staff, there are not a lot of good things that happen to it. Best case, it is separated from the fermentation broth and somehow disposed. This can be as benign as being sent to landfill or slow conversion by anaerobic digestion (equivalent of aging gracefully). The problem comes when the company wants what is inside the cell walls. This usually involves a process called disruption, which is just a polite term for blowing the microbe to pieces. The worst part is that the microbe will not likely have enough time at the company for their stock options to fully vest.

Management does not listen to my input – as microbes cannot communicate to management through the normal channels (Power Point presentation with complex drop-ins), they do it by less formal, but highly effective means. To show their value to the organization, they often hold any major advancement until days before a board meeting or investor call to make management sweat about how they are going to explain the shortfall, causing a traditional rush to get last minute results ready for prime time.

The stress is unbearable – While not intended to harm the organism, stress is a key part of getting desired outcomes. The microbe flourishes through the growth stage and before you know it, the company starts messing with it by taking away key nutrients or somehow shocking it. This is not done out of spite, but because these triggers cause a change in behavior to a more favorable state for the company. Sounds reasonable, unless you are the one they are shocking.

My goals are always changing – How is a microbe expected to know what is needed from them when the goalposts keep changing. One week, the priority is yield on sugar, the next week ultimate fermentation titer and then the invasive questions on whether it meets the definition of "organic" or "natural" begin. If a microbe is trying to do its part to support company goals, it is important to have them clearly outlined. Good thing microbes don't get an incentive bonus, as the evaluation criteria change daily.

Does someone really need to count beans – It is hard for a microbe to hear the technical staff through the sound of bubbling gases in the fermenter, but the one common theme they hear is that a primary source of concern at the company starts with the "bean counters". What confuses the microbe (as a simple undeveloped organism, incapable of higher-level thinking) is why the company needs someone counting beans. Worth pondering.

As with all key employees, it is critical to understand how the selected microbe views your organization to make sure you do not find them working for your competitor.

KEY TERMS

KEY TERMS

The following is a summary of key terms within the handbook and their practical definition.

Aerobic Fermentation - Fermentation that requires oxygen and involves sparging of air.

Agitated Fermenter - Fermentation tank with mechanical agitation, usual aspect ratio of 2.5 – 3.5 with air injected at the bottom.

Airlift Fermenter - Fermentation tank with agitation provided by rising air bubbles, has internal structure to direct flow. Typical aspect ratio of 6 – 7. Often confused with bubble column that does not have internal structure.

Anaerobic Fermentation - Fermentation that is done in the absence of oxygen. Common example is corn ethanol.

Appropriations Grade Cost Estimate - Cost estimate based on FEL-3 engineering study, accuracy of +10%/-10%. Required for debt funding approval.

Aseptic Fermentation - Fermentation process where only a single target organism is expected at the end of the fermentation.

Aspect Ratio - Ratio of height of tank divided by diameter of tank. Larger the number, the taller and thinner the tank is.

Binding Equipment Estimate - Firm proposal from vendor with binding price, delivery time and contractual terms, ready for contractual execution.

Bubble Column Fermenter - Fermentation tank with agitation provided by rising bubbles with no internal structure. Typical aspect ratio of 6 – 7.

Budgetary Equipment Estimate - Non-binding cost estimate from vendor outlining price, delivery time and key contractual terms for equipment purchase.

CAPEX - Capital cost for design and construction of a proposed process facility.

CIP - Clean-in-place, protocol used to clean equipment between operations, usually done with 60C sodium hydroxide solution.

Clarification - Process of separating solids from liquids.

CMO - Contract Manufacturing Organization, a facility that can be contracted to run a manufacturing process in their equipment to make the proposed product.

Commercial Plant - Full-scale facility that is of an adequate size to be independently profitable.

Commissioning - Stage in a new facility when the plant has been built and is going through initial testing to confirm equipment works, but is not ready to make product.

Construction - Physical assembly of the purchased equipment into a facility.

Cost Reimbursable - Contracting mechanism for engineering and construction where owner manages the work and pays for effort expended. Synonymous with time and materials.

Demo Plant - Demonstration scale facility – built of a size to demonstrate the process adequately to justify a commercial scale facility, but is not typically profitable.

Detailed Design - Detailed engineering phase where FEL concepts are turned into detailed construction drawings and specifications to support facility construction.

Disruption - Unit operation where microbial cells are partially or completely lysed to gain access to materials inside the cell wall.

DMG - Di-Methyl Glop – mythical chemical compound used as a generic example material in process examples.

Drying - Removing water from a solution to produce a final dry product.

DSP - Down Stream Processing – purification and refining unit operations to turn fermentation broth into final product.

EPC - Engineer, procure, construct – describes the process of taking a plant from early stage development to a constructed facility. Infers contractor responsibility for project.

EPCM - Engineer, procure, construction management – Similar to EPC, but with owner maintaining active project control.

Feasibility Level Cost Estimate - FEL-2 level cost estimate, +25%/-15% level of accuracy.

FEED - Front End Engineering Design – describes the complete FEL process from FEL-1 through FEL-3. Stage-gate process used for process project development.

FEL-1 - Front-end loading level 1, concept screening level of definition.

FEL-2 - Front-end loading level 2, feasibility study level of definition.

FEL-3 - Front-end loading level 3, project definition level of definition.

Fermentation - Biologic process using microbes to convert a carbon source into products.

Fermentation Broth - Liquid product of fermentation process that contains cells, media and target product before purification.

Fixed Price - Engineering and construction approach where the contracting firm is able to provide a fixed price for the work outlined.

FSMA - Food Safety Modernization Act – US governmental regulation covering how food products are manufactured.

HTST - High Temperature Short Time – a packaged process system used to sterilize liquid streams for fermentation.

Independent Engineer - Engineer hired by investors or funding source to provide an assessment of technology or advise of viability of a project.

Industrial Biotechnology - Biotechnology focused at making foods, fuels and chemicals.

Integrated Operations - Manner of operating a proposed process, involves running all unit operations simultaneously to represent the planned commercial process.

Mechanical Completion - Point of a process construction project when construction is complete, but none of the equipment has been tested.

Media - Salts, minerals and vitamins added to the fermentation process to support cell growth.

Microbe - Microbial organism.

Modularization - Construction process where large portions of the facility are built in sections and assembled at the site.

Novel Food - New or unique food product, made from industrial biotechnology processes.

OPEX - Operating expense, synonymous with manufacturing cost.

Order of Magnitude Cost Estimate - Cost estimate based on FEL-1 engineering study, accuracy of +40%/-25%.

Owners Engineer - Engineer or consultant that works for a company commercializing a technology and acts as advisor and interfaces with outside engineering and construction firms.

P&ID - Process and Instrumentation Diagram – detailed engineering drawings that outline the core details of how a process will be designed, operated and controlled.

Pharmaceutical - Regulated products synonymous with drugs, often referred to as active pharmaceutical ingredients (APIs).

Pilot Plant - Research level facility used to prove out process and make small samples for testing.

Project Wrap - Cost and performance guarantees that the EPC firm contractually agrees to warranty the project for.

Proof of Concept - When the basic science of a proposed process is demonstrated.

Purification - Converting raw crude intermediate process streams into a final product.

Sanitization - Uses in food applications, cleaning surfaces to be free of harmful organisms.

Scale-up Factor - Ratio of the proposed facility size to the current facility size.

SIP - Steam-in-place – method of sterilization for aseptic fermentation equipment.

Skid-Mounted - Equipment that comes in a complete operable package, that can be shipped by truck and placed directly onto a flat surface.

Start-Up - Point in bringing a new facility on line where systems have been tested and raw materials are used to try and make product.

Sterile Filter - Filter of 0.2 micron that is used as a boundary to keep contamination out of fermentation from liquid and gas streams entering.

Stick-Built - Method of constructing a process facility where the bulk of the fabrication work is done on the site.

Unit Operations - Chemical and physical processes for producing and refining chemical products.

Made in the USA
Las Vegas, NV
06 June 2022

49870115R00059